# Next Generation Digital Communication Systems

# Next Generation Digital Communication Systems

Edited by **Glynn Clermont**

**C**WILLFORD PRESS

New York

Published by Willford Press,
118-35 Queens Blvd., Suite 400,
Forest Hills, NY 11375, USA
www.willfordpress.com

**Next Generation Digital Communication Systems**
Edited by Glynn Clermont

International Standard Book Number: 978-1-68285-027-5 (Hardback)

Printed in the United States of America.

# Contents

# Preface

Digital communication systems have revolutionized the telecommunications industry in the present age. They have become an integral part of the corporate & business domains as well as different individual groups or consumer circles. There is a rising demand for conceptual and practical advancements in the field of digital communication systems. This extensive book encompasses some of the most significant concepts in digital communication that include mobile and wireless communication systems; frameworks, tools & techniques applied in digital communication systems; evaluation and application of digital communication systems in different spheres, etc. It contains researches and case-studies by experts and academicians from diverse regions of the world that discuss the latest improvements as well as innovations which would contribute to the overall development of next generation digital communication systems.

Various studies have approached the subject by analyzing it with a single perspective, but the present book provides diverse methodologies and techniques to address this field. This book contains theories and applications needed for understanding the subject from different perspectives. The aim is to keep the readers informed about the progress in the field; therefore, the contributions were carefully examined to compile novel researches by specialists from across the globe.

Indeed, the job of the editor is the most crucial and challenging in compiling all chapters into a single book. In the end, I would extend my sincere thanks to the chapter authors for their profound work. I am also thankful for the support provided by my family and colleagues during the compilation of this book.

Editor

# A methodology to counter DoS attacks in mobile IP communication

Sazia Parvin[a,*], Farookh Khadeer Hussain[b] and Sohrab Ali[c]
[a]*Digital Ecosystems and Business Intelligence Institute, Curtin University, Perth, Australia*
[b]*School of Software, Faculty of Engineering and Information Technology, University of Technology, Sydney, Australia*
[c]*The People's University of Bangladesh, Dhaka, Bangladesh*

**Abstract.** Similar to wired communication, Mobile IP communication is susceptible to various kinds of attacks. Of these attacks, Denial of Service (DoS) attack is considered as a great threat to mobile IP communication. The number of approaches hitherto proposed to prevent DoS attack in the area of mobile IP communication is much less compared to those for the wired domain and mobile ad hoc networks. In this work, the effects of Denial of Service attack on mobile IP communication are analyzed in detail. We propose to use packet filtering techniques that work in different domains and base stations of mobile IP communication to detect suspicious packets and to improve the performance. If any packet contains a spoofed IP address which is created by DoS attackers, the proposed scheme can detect this and then filter the suspected packet. The proposed system can mitigate the effect of Denial of Service (DoS) attack by applying three methods: (i) by filtering in the domain periphery router (ii) by filtering in the base station and (iii) by queue monitoring at the vulnerable points of base-station node. We evaluate the performance of our proposed scheme using the network simulator NS-2. The results indicate that the proposed scheme is able to minimize the effects of Denial of Service attacks and improve the performance of mobile IP communication.

Keywords: Security, communication, mobile IP, DoS, DDoS, attack, filtering, monitoring

## 1. Introduction

In the last few years, demands for mobile computing have increased noticeably with the increasing use of powerful hand-held communicating wireless devices such as cell phones, i-pod, PDAs, laptops etc. Remarkable progress in the areas of mobile computing has streamlined our communication. In spite of having worldwide Internet access, it is not expectable to reap all the benefits until secure communication over the communication channel is ensured. There are different kinds of security threats in current wireless networks. Palmieri et al. [35] proposed a new active third party authentication, authorization and security assessment strategy for next generation wireless mobile networks. Algorithms are proposed in [15] to quantify the trust between all parties of 3G-WLAN including cellular and IP technologies integrated networks to secure user authentication in wireless networks.

Providing security for Mobile IP communication is a complex process because of its inherent dynamic characteristics like frequent changes in the point of attachment, absence of central administration etc. Various kinds of attacks on Mobile IP Communication can disrupt normal functioning. Among all these attacks, Denial of Service (DoS) is a difficult problem and becomes an increasing threat to the current Internet [34].

---

*Corresponding author. E-mail: sazia.parvin@postgrad.curtin.edu.au.

The main target of Denial of service (DoS) attack is to consume a server's resources such as network bandwidth, computing power, main memory, disk bandwidth etc. This results in either limited or no resources being left to process further upcoming requests from legitimate clients. Thus, a DoS attack can paralyze any user, site or server in the Internet. There is another category of DoS attack which is known as Distributed Denial of Service (DDoS) attack, where many malicious users simultaneously attack the same site/server. As a result, the impact of the attack on that node is much more severe. Therefore, DDoS is often considered as one of the worst possible attacks in the current Internet [34]. DoS might be a great threat in the ubiquitous medical sensor networks [14] by changing the patient's data. Delot et al. [11] proposed a data management solution for event exchange in vehicular networks to avoid dangerous/undesirable situations where DoS might be a great threat. A general solution is proposed to protect the system against DoS attack by filtering DoS attack requests at the possible earliest point before they use much of the server's resources [31]. A considerable amount of work has been done to prevent DoS attacks on wired networks. In contrast, however, there has been little research focused on Mobile AdHoc Networking. However, DoS attack is now considered to be a great threat to Mobile IP communication. In this work, we have proposed a new technique to prevent DoS attack in Mobile IP communication. We propose two techniques in this paper: (i) packet filtering technique in the domain periphery router and base station for discarding the suspected packets sent by the DoS attackers by (ii) queue monitoring technique to mitigate the effect and damage by DDoS attack in the vulnerable points of the network.

### 1.1. Objectives and motivation

Today's world is enjoying tremendous advantages thanks to the advancement in the area of mobile computing. Mobile IP uses one of the technology which can support of various mobile data and wireless networking related applications. Users can enjoy abundant seamless roaming and fast feasible application anywhere at any time by using mobile IP. Moreover, users can get continuous access to the Internet through Mobile IP. Hence, ensuring security services for Mobile IP has now become a major challenging issue of researchers. There are different kinds of attacks in Mobile IP communication which can disrupt its normal communication. Denial of Service (DoS) attack is one of the greatest threats to today's Internet [34]. A DoS attack is capable of harming any user or server connected to the Internet. You et al. [49] proposed fast mobile IPv6 security by minimizing an involvement of the authentication server. A huge amount of work [11,17,21,25] has been done to prevent or mitigate this type of attack. But most of this work has been done for wired communication, while some relates to mobile adhoc Networking. Providing security mechanisms against DoS attack for Mobile IP communication is an outstanding research issue. Although some works have enhanced the security for Mobile IP communication, most provide only a general solution and do not address the problem clearly. They do not provide the security requirements of the applications and do not adequately address specific attacks. Therefore, in this paper, we propose a solution to counter DoS attack in Mobile IP communication.

### 1.2. Problem statement

Recently, Mobile IP has received closer attention in areas of both research and application. The application of Mobile IP is increasing day-by-day. Mobile IP devices are vulnerable to attacks because of their dynamic nature. Most of the current researchers pay attention how to increase the network efficiency in mobile IP Environment. A great deal of research has been done to overcome this problem [33,39, 46]. Therefore, there are quite a number of proposals proposed in [33,39,46] which can show high performance through route optimization. Providing security in Mobile IP using the firewall concept has been proposed in many research proposals. That is, the Mobile IP users can securely access a firewall-protected network. Much work has been done to provide security to Mobile IP using IP Sec. Original Mobile IP protocol provides some general security services which gives some protection to mobile IP communication. However, there still exist numerous threats that cannot be prevented by current general security services. Much work has been done to remove the inefficiencies of existing security protocols for mobile IP communication, but little has been done to protect mobile IP communication from specific attacks like DoS attack [4]. It is difficult to mitigate DoS attack in Mobile IP communication due to the following problems:

- A very little research has been done to mitigate DoS attacks in the Mobile IP communication. Little research has been carried out to compare and categorize different types of DoS attacks and find defense mechanisms.
- There is no effective defense mechanism against DoS attack in Mobile IP communication as well as no guideline for the selection of defense mechanism.
- Existing solutions and defense mechanisms have been evaluated based on assumptions and according to limited criteria.

### 1.3. Aim of this work

The main goal of this work was to analyze the effect of Denial of Service attack on mobile IP communication and provide a reliable solution for ensuring security against this attack. Our aim is to use a packet filtering technique for discarding the suspected packets sent by the DoS attackers. We are also interested in monitoring the queues in the vulnerable points of the network to mitigate the effect and damage caused by DDoS attack.

### 1.4. Scope of this work

There are different ways to ensure security against Denial of Service attack. We can detect the suspected packets related to DoS attack and discard those packets before the attack occurs. We can do this at the important points of the network. Sometimes we cannot detect the attacks before they occur. In these cases, we have to mitigate the amount of damage done during the attack. As a precaution, we can also mark every packet going through the network routers by keeping track of its previous source and next destination. When a packet is detected as a suspicious packet, we can trace it back to its original source following the marked paths. Thus, we can detect the real attacker. Possible approaches for ensuring security against Denial of Service attack are as follows:

1. DoS attack detection and prevention
2. Mitigating the amount of damage while being attacked
3. Back-tracking and packet marking for detecting the attacker

Our work focuses on the first two approaches. The third approach has been omitted here.

## 2. Major contributions

The major contributions of this paper is to analyze the effects of Denial of Service attack on mobile IP communication in detail. We propose here packet filtering techniques in different domains and base stations of mobile IP communication to detect suspicious packets and by using these techniques we can improve the performance. If any packet contains a spoofed IP address which is created by DoS attackers, the proposed scheme can detect this and then filter the suspected packet. The main contribution of this paper is:

- We filter the domain periphery router and base station to mitigate the effect of Denial of Service (DoS) attack.
- We monitor the queue at the vulnerable points of base-station node to minimize the Denial of Service attack.
- We evaluate the performance of our proposed scheme using the network simulator NS-2 and show that our proposed scheme is able to minimize the effects of Denial of Service attacks and improve the performance of mobile IP communication.

## 3. Related works

Mobile IP produces new security threats for wireless networks and there is a comparatively higher chance being attacked by hostile opponents in comparison to wired network due to their wireless media. In the following sections we highlight different works related to ensuring the security of mobile IP communication.

### 3.1. Sec MIP

Torsten Braun and Marc Danzeisen [5] proposed Secure Mobile IP (SecMIP) to provide security to Mobile IP using IP Sec. In their work [5], they proposed the way in which a Mobile node access the network securely. Each Mobile Node needs to authenticate itself using IP Sec if it wants to traverse the firewall. Therefore, a secure IP Sec tunnel is established between the firewall and the Mobile Node for the communication. The functionalities of this tunnel and working principles are fully described step by step in this paper [5]. SecMIP is implemented and a performance result is also presented in this work. The main advantage of this solution is that it does not require the introduction of any new protocols or the modification of any existing protocol. But the problem of this proposition is that if a Mobile Node wishing to transfer data to a correspondent node, it must send it via the Home agent [5]. So, there is a chance of traffic increment in the network as a MN cannot securely transfer data to a correspondent node directly. The authors in [19] proposed an effective method for measuring the accuracy of IP multicast based multimedia transmission and the main focus of this method is to find the accuracy and the complexity of the user model describing user movement in the network.

### 3.2. Use of IP Sec in mobile IP

John K. Zao and Matt Condel [51] used IP Sec ESP protocol in Mobile IP to defend against both passive and active attacks. In this approach, the security requirements are fulfilled by establishing a MIP-IPSec tunnel between MN-HA, HA-FA and FA-MN. Some modifications have been proposed to Agent advertisement and to the registration of request messages. The authors in [29] proposed a Diffie-Hellman key based authentication scheme that utilizes the low layer signaling to exchange Diffie-Hellman variables and allows mobility service provisioning entities to exchange mobile node's profile and ongoing sessions securely.

## 3.3. Secure mobile networking

Vipul Gupta and Gabriel Montenegro [22] proposed some enhancements to the basic MobileIP protocol so that authorized users can access a network that is protected by some combinations of source filtering routers or the network such as firewalls, which are using private address space for security reasons. The use of private addresses is quite challenging to Mobile IP because the Mobile node cannot use its home address to communicate with a correspondent node while outside its protected network. The concept of a secured Mobile IP presented in this paper [22] is able to solve the security problems of Mobile IP in an efficient and successful way, but it requires the introduction of new protocols. The authors in [10] proposed a novel secure and efficient ID-based mobile IP registration protocol in Mobile IP networks by considering performance while providing the security. The authors in [30] proposed a general analytical model through Authentication, Authorization, and Accounting (AAA) framework architecture which is able to protect incoming messages from malicious attackers in Mobile IP. The authors in [42] proposed an agent-based model to find the issues of intrusion detection in cluster based mobile wireless ad hoc network environment and evaluated how agent based approach facilitates flexible and adaptable security services. Ciou et al. [9] proposed an effective and efficient handover key management and authentication scheme during handover by speeding up the handover process through Diffie-Hellman key exchange scheme in order to increase the security level for mobile stations (MSs).

## 3.4. Secure mobile IP protocol

Atsuhi Inoue, Masahiro Ishiyama, Atsushi Fukumoto and Toshio Okamoto [26] proposed to modify Mobile IP protocol with IP Sec. Datagram entering the network and exiting the visiting network both are securely processed using IP Sec. Here, secure Mobile IP is implemented on gateway servers and Mobile Nodes for evaluation purpose. In this approach, a mobile node with IP Sec processing capability is called a Secure Mobile Node (SMN) and a firewall with IP Sec processing capability is called a Security Gateway (SGW). It is recommended to place SGW between the inside network and the Internet. But the organization can also place SGW inside it's network to protect of the specific division of the network. A dynamic gateway discovery has beeb proposed to select the best location for a specific SGW and also a communication model showed the details for traversing of SGW using IP Sec.

## 3.5. Securing binding update message

A new protocol is proposed to secure binding update messages and protect against redirect attacks [13]. This protocol uses public key cryptosystems which is based on digital signature and Diffie-Hellman key exchange algorithm. Chang et al. [7] designed the proper technical measures ('Software Tamper Resistance') for mobile game service to minimize its vulnerability by using cryptography techniques. There are two existing protocols proposed by the IETF Mobile IP working group. They analyzed the weakness of these existing protocols and generated the report. The authors in [7] mentioned secure key negotiation as a challenging issue for the successful deployment of mobile IPv6. In their work, they analyzed and showed the performance of optimal binding-management-key refresh interval in mobile IPv6 networks. Chen et al. [8] a novel authentication scheme to ensure the security of the transmitted messages against known attacks by integrating fingerprint biometrics, related cryptology and a hash function mechanism and this scheme requires a low implementation cost.

## 3.6. Mobile IP security system

John Zao, Stephen Kent, Joshua Gahm, Gregory Troxel, Matthew Con-dell, Pam Helinek, Nina Yuan and Isidro Castineyra [50] designed and implemented Mobile IP Security System (MoIPS) using a public key system for route optimization. According to their approach, all the hosts and mobile agents must use MoIPS certificates with a view to authenticate the Mobile IP control messages. Public key technology can efficiently meet the requirements of key management. It is mentioned in [27] that the X.509 PKI uses two types of certificates: Certificate of Mobile IP Control Message Authentication (MoIPS Certificates) and Certificates for IP Security Services (IP Sec Certificates). There are many solutions for ensuring the security in Mobile IP. Most of the solutions are general and do not focus on specific attacks. Some research has also been done on the detection and prevention of DoS attacks, but most of them are for either wired communication or mobile ad hoc networking. We discuss several of these works below.

### 3.7. DoS attack detection approaches

Most of detection approaches rely on finding the malicious party who has created a DoS attack and subsequently made damage [23]. However, identifying and tracking the real attacker is not a easy task. The authors in [23] mentioned two possible reasons: (i) the attacker spoofs the source IP address of the attacking packets and (ii) the Internet is stateless, which makes a sense that whenever a packet goes through a router, the router does not store any information (or traces) about that packet.

#### IP trace back
Yang Xiang and Wanlei Zhou [47] proposed a protection system against DoS attack by a large scale IP Traceback method. They named their system FDPM (Flexible Deterministic Packet Marking). IP Traceback can identify IP packets of their sources without any prior information about the source address field of the IP header. This technique is extremely beneficial for identifying the sources of the attackers and taking appropriate actions against them. This proposition is applicable for wired networks.

#### ICMP trace back
Bellovin [2] proposed the idea of ICMP Traceback messages. In this approach, every router use one sampling technique for each forwarded packets with a very low probability and reports to the base station by sending an ICMP Traceback message while packets are passed through the network from the attacker to the victim. An ICMP Traceback message carries the information about the previous and next hop addresses of the router, timestamp, portion of the traced packet, and authentication information [2]. The most challenging thing of this approach is that the attacker makes the victim confused by sending many false ICMP Traceback messages. Barros [1] proposed a modification to the ICMP Traceback messages to figure out the Distributed DoS (DDoS) attacks by reflector.

#### Packet marking
Burch and Cheswick [3] proposed to discover some path information into the header of the packets instead of routers sending separate messages for the sampled packets. This marking technique is deterministic or probabilistic. In the deterministic marking technique, all packets are marked by every router marks. The major drawback of the deterministic packet marking is that the packet header increment is proportional to the number of hops increment on the path. In probabilistic packet marking (PPM) technique, it encodes the path information of every packet into a small possible fraction. It is assumed that a huge amount of packet traffic rush towards the victim during a flooding attack. Therefore, many of these packets will be marked at routers by packet marking technique during their travel from the source to the victim. The marked packets which contains enough information will be able to provide enough information to trace the network path from the victim to the source during the network travel.

## 3.8. DoS attack prevention approaches

We discuss here some DoS attack preventive approaches. The goal of these approaches are to identify the attacked packets and discard them from the network before they attack the victim. We examine several packet filtering approaches [10,12,36,48] that are able to detect the attack packet and discard them from the network.

### Reputation based incentive scheme

Denko et al. [12] proposed a DoS attack prevention scheme using a reputation-based incentive scheme which works fine in mobile ad hoc networks. With this proposed mechanism, the reputation of all nodes in this network will be updated based on their behavior (good or malicious). They examined their technique on both active and passive DoS attack and claimed their technique will motivate the nodes in the ad hoc network to prevent both types of DoS attacks.

### DoS limiting network architecture

TVA system, Packet Passport system and StopIt system are proposed in [48] for limiting DoS attack in wired communication. Traffic Validation Architecture (TVA) is short-term authorization technique. In this TVA technique, senders will stamp their authorization on received packets whenever they will receive from receivers. The Packet Passport system is a piece of authentication information embedded into an IP packet that authenticates the source IP address. StopIt is a packet filtering system that blocks the undesired traffic it receives.

### Ingress filtering

Ingress routers can filter the incoming packets to a network domain. These filters has the capability to verify the identity of packets entering into the domain. Farguson and Senie [10] proposed Ingress filtering which can filter the incoming packets by dropping any traffic with an IP address that does not match a domain prefix connected to the ingress router.

### Egress filtering

The concept behind an egress filter [6] is that only packets with one's network source address should be leaving one's network. In computer networking, egress filtering is a technique which can monitor and restrict the flow of information from one network to another. Router or firewall of a networks can always examine the TCP/IP packets that are being sent out of the internal. Packets that do not meet security policies, are denied by the egress filter. Egress filtering ensures the presence of the unauthorized or malicious traffic in the internal network. The idea is to permit only those packets from trusted hosts to leave your network [18].

### Route-based filtering

Route-based filtering is proposed by Park and Lee [36] which can filter out spoofed IP packets based on route information. In this system, every router keeps track of the incoming and outgoing paths of each packet. When a packet is detected as malicious, then its route from sender to receiver is marked as dangerous. In future, packets coming from that route would be blocked. Qadri et al. [41] mentioned that the lack of centralized routing and network resource management becomes an obstacle fore video streaming within a VANET.

As we can see from a comprehensive review of the existing literature review, we represent a meta-literature analysis in the Table 1:

Table 1
Literature review

| Packet filtering techniques | Proposed schemes | Outcomes | Shortcomings |
|---|---|---|---|
| Reputation Based Incentive Mechanism [12] | Reputation Based Incentive Scheme | They proposed reputation-based incentive mechanism for encouraging nodes to cooperate both in resource utilization and preventing DoS attacks [12] | The approach is only used for clustering architecture in MANETs in a localized or distributed manner |
| Traffic validation Architecture [48] | Traffic validation Architecture | They proposed the design and evaluation of TVA, a network architecture that limits the impact of Denial of Service (DoS) floods from the outset | The approach works for wired communication |
| Ingress filtering [10] | Ingress filtering techniques | This mechanism can drop any traffic with an IP address that does not match a domain prefix connected to ingress router [10] | This approach works for Internet Service Providers and the Internet community |
| Egress filtering [6] | Egress filtering technique | Packets that do not meet security policies are now allowed to leave-they are denied "egress" [6] | This approach is only to permit those packets from trusted hosts to leave own network [18]. |
| Route based filtering technique [36] | Route-based distributed packet filtering technique [36] | Every router keeps track of the incoming and outgoing paths of each packet. | This approach is suitable for Internet-based system. |

Clearly, from a comprehensive review of the existing body of literature and the meta-literature analysis on mobile IP, we can conclude that (a) there has been significant research and advances made in the area of mobile IP security (b) there is no-method to counter the DoS attacks in Mobile IP.

## 4. Methodology to counter DoS attack

It is too complex to detect and prevent DoS attack in a dynamic network such as wireless sensor networks, mobile ip networks. So, its easier to divide a large network into small and manageable groups and implement security mechanisms in each group in a distributed manner instead of thinking the whole network [37]. For this purpose, here we follow the same hierarchical architecture which we already proposed in [37]. More specifically, we propose to first divide the network into domains. Then each domain can be divided into clusters, each of which consists of one or more wired or mobile nodes or base station. We argue that, if we can manage and secure each cluster efficiently, we can in turn secure each domain and thus the whole network becomes secure. This is basically a distributed approach where each domain or cluster is independent.

Clustering architecture is a distributed and scalable architecture for monitoring and securing the networks.

There are other benefits of clustering architecture as stated in [12]. Clustering architecture can monitor the network continuously and detect the attack in the network. After that, it provides prevention mechanism based on the attacker's characteristics. So, this architecture can minimize the network bandwidth by reducing storage and communication overhead. The Fig. 1 shows the clustering architecture of a network:

Domain 0 (2 cluster each with 1 node)

Cluster 0 W(0) 0.0.0

Cluster 1 W(1) 0.1.0

Domain 1 (1 cluster, 2 node)

HA 1.0.0

Cluster 0 MN 1.0.1

Moving

Domain 2 (1 cluster, 1 node)

FA 2.0.0

Cluster 0

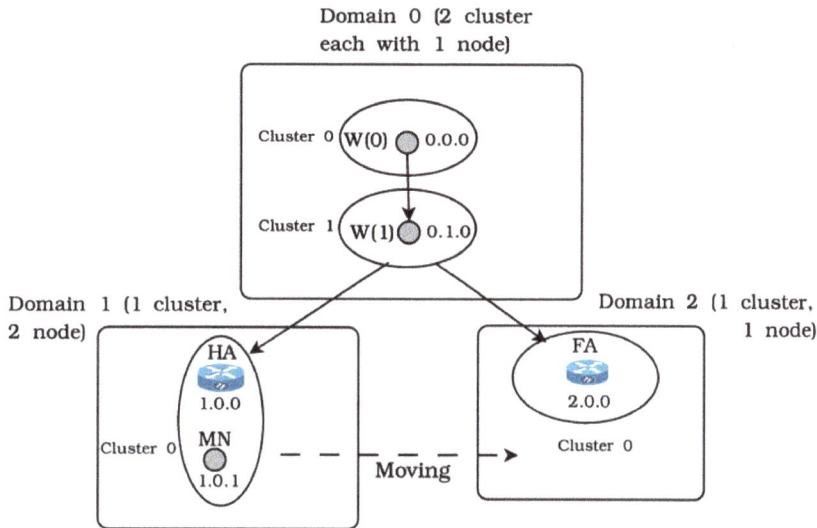

Fig. 1. Clustering architecture of mobile IP communication.

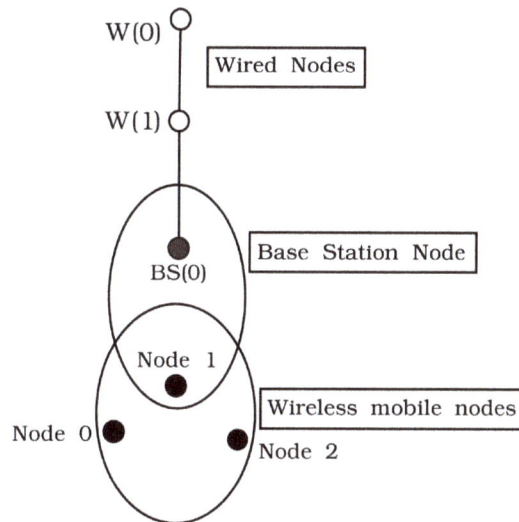

W(0)

Wired Nodes

W(1)

BS(0)

Base Station Node

Node 1

Node 0

Wireless mobile nodes

Node 2

Fig. 2. Topology for wired cum wireless scenario.

Figure 2 shows a wired-cum-wireless topology through which we can exchange packets between a wired and wireless domain through using a base-station. But at any time a mobile node can roam outside the domain of its base-station and be still in a communication link so that it can receive any packets which is destined for it [12]. That is why we have extended the Mobile IP support in this wired-cum-wireless scenario.

Figure 3 shows a wired domain. This domain has 2 wired nodes, named as W0 and W1. In this architecture, We have 2 base-station nodes, Home Agent (HA) and Foreign Agent (FA) respectively. In this architecture, W1 is connected to HA and FA. There is a roaming mobile node called Mobile Node (MN) that moves between the communication range of its home agent (HA) and foreign agents (FA). A TCP flow will be set up between any node (e.g. W0) and MN. According to the architecture, when a MN

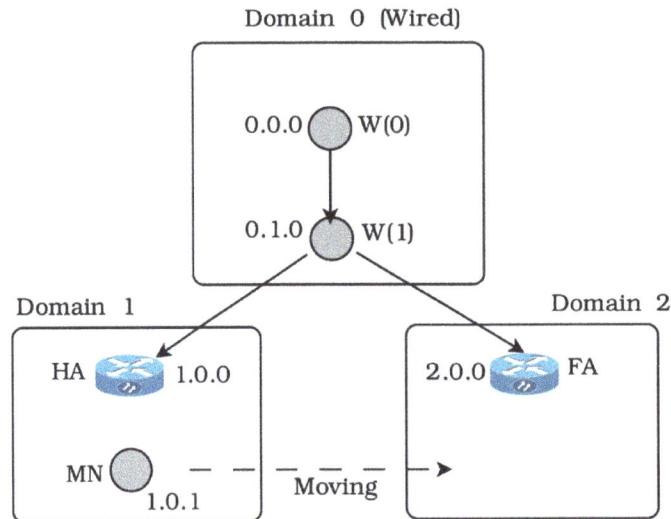

Fig. 3. Topology for Mobile IP simulation.

moves out from the domain of its HA, into the domain of FA, if any packet is sent to the MN during this time interval, the packet will be automatically redirected to the FA by its HA as per definitions of mobile IP.

In the above architecture, we have set up one as wired domain (denoted by 0) and 2 as wireless domains (denoted by 1 and 2 respectively). According to the clustering architecture shown above, the addresses of two wired nodes are 0.0.0 and 0.1.0. The first section of the address represents the domain number, the second section represents the cluster number and the third section is the node number. In the first wireless domain (domain 1) we have a base-station, HA and mobile node MN, in one single cluster. So their addresses are represented as 1.0.0 and 1.0.1 respectively. And for the second wireless domain (domain 2) we have a base-station, FA with an address of 2.0.0. However, when the MN moves into the domain of FA, the packets originating from a wired domain and destined for MN will reach to as per definitions of Mobile IP protocol. The above figure depicts a basic structure of a Mobile IP communication network which may have a huge number of domains. Each domain may contain a different number of clusters and each cluster can contain a different number of nodes.

We present here the following advantages of our proposed hierarchical structure to securing the network.

First, this approach is scalable, because here we are dividing the whole network into several sub-networks and managing and securing them separately.

Second, each domain or cluster can be controlled independently. That means that the security mechanisms in a cluster lying in the more dynamic portion of a large network may not match that of a cluster lying in a comparatively static portion.

Third, the detection of an attacker becomes easier in clustering architecture as it provides localized information. On the other hand, attacker detection is much more complex in a large network with a flat hierarchy.

Fourth, this localized and distributed feature reduces storage, processing and communication overheads, thereby optimizing network bandwidth utilization.

After dividing a large network into a hierarchical structure, we propose three types of filtering techniques. These are:

1. Filtering in the Domain Periphery Router
2. Filtering in the Base-station
3. Queue Monitoring in Base-station Node

Each of these techniques is described in detail below.

### 4.1. Filtering in the domain periphery router

According to [37], there is an edge or periphery router in each domain, through which each packet within the domain has to pass when going to another domain. In this proposed method, we propose filtering technique in the domain periphery router with a view to filter the malicious packets. Basically, home agents or foreign agents act as the periphery routers in a domain. According to [37], when a mobile node is in a foreign network and wants to register with its home agent via its foreign agent, then the foreign agent would keep track of the addresses associated with that mobile node using a caching mechanism. After registration, when the mobile node wants to attack any node outside its current domain by spoofing the source address, then the domain periphery router will detect and filter that packet as the spoofed address is unknown to itself. When the mobile node tries to attack from its home network, then the home agent will block the packet which contains addresses outside of this network or which is unknown to the home agent. Here we use the concept of egress filtering because egress filtering is used to filter the packets leaving a network [6].

**IF** packet's source address is within domain's address
**THEN** forwards the packet
**ELSE** discard the packet

### 4.2. Filtering in the base station node

According to our proposal in [37], If the attacker resides inside the same domain of the victim, then the edge or periphery router could not detect the attacking packet. That's why we have proposed an additional filtering technique in the Base Station node (HA or FA) to which the mobile nodes are connected. Basically the base station nodes (HA or FA) in mobile IP communication are the main targets of the attackers, because the mobile nodes get services from these base stations. So detecting and preventing attacks in the base station nodes is very important. In our proposed scheme the base station node will filter a packet if one of the following events occurs as described in [37]:

- If the base station's router queue overflows
- If there are many packets from same domain or same cluster, because the attacker nodes at first take help from the neighbor for attacking any target
- If most of the bandwidth of the network is consumed by DoS attackers, then the network will be congested. If the network gets congested then incoming packets should be discarded for the time beings

### 4.3. Queue monitoring in base station node

This scheme is for preventing Distributed Denial of Service (DDoS) attack. When a node of any domain is under DDoS attack, its corresponding base-station will be overloaded by the attacking packets. Most of the resources of the base-station will be consumed by the attacking packets. As a result, fair nodes will be barred from services. In this case, we propose to monitor the queue and define its size for setting up in an adaptive manner. More specifically, we consider here the case of sudden increase of tiny packets (like TCP SYN packets) in the queue. It is stated in [38] that a sudden increase of such packets raises the probability of a Denial of Service attack. So, dropping packets in this case would reduce the effect of the attack. In this way we can save the base-station node and its mobile nodes from DDoS attack. Here we use the concept of ingress filtering because ingress filtering is used to filter the packets entering a network.

We propose to regularly measure the queue on the basis of small (near to size of TCP SYN packets) packets at the base-station nodes. Then we set up a predefined level of the queue size. When the queue size of the packets increases to above a predefined level, the queue size will be limited by an adaptive value considering other bigger size packets in the queue. This indicates that we need to filter only the suspicious packets, not any one of the data packets.

As described earlier in [37], the problem with this approach is that some applications use small packets, which may also be dropped during this filtering process. To save these packets, we propose to drop packets in case of a sudden and large increase of small packets in the queue. We argue that the applications that use small packets do not send packets at a very high rate.

## 5. Simulation and result analysis

Network Simulator 2 [16,20], is used as the simulation tool in this paper. We have created simulation environments and conducted a performance evaluation using the network simulator NS-2 [16,20], as it provides the wide range of features and it has an open source code that can be modified and extended. We have created a wired-cum-wireless topology through which we can exchange packets between a wired and wireless domain via a base-station. But a mobile node may roam outside the domain of its base-station at any time and should still continue to receive packets which is destined for it. That is why we have implemented the Mobile IP protocol with that wired-cum-wireless scenario. At first we have implemented the DoS attack scenario without protection in this Mobile IP environment. After that, we have simulated the scenario by applying filtering technique in the domain periphery router and in the base station node. Finally, we simulated the queue monitoring in the base-station node as a part of our proposed scheme. Then we compared the performance of the simulation results.

## 6. Description of the simulations

We have undertaken all the simulations required for this work, dividing this process into four steps:

1. Creating a wired-cum-wireless scenario
2. Running Mobile IP in the wired-cum-wireless topology
3. Implementation of mobile IP communication with Denial of Service attack
4. Implementation of mobile IP communication with proposed technique

All of these steps are described below:

### 6.1. Creating a wired-cum-wireless scenario

In this section, we have simulated a mixed scenario consisting of a wireless and a wired domain. In this scenario, data is exchanged between the mobile and non-mobile nodes. As we set up a mixed scenario, so we have 2 wired nodes, W(0) and W(1), connected to our wireless domain consisting of 3 mobile nodes (nodes 0, 1 and 2) via a base-station node, BS. Here, base-station nodes actually act as gateways between wireless and wired domains and allow packets to be exchanged between the two types of nodes. Figure 2 shows the topology for this example.

The DSDV Ad hoc routing protocol is used here. Also, we defined TCP and CBR connections between the wired and wireless nodes. For mixed simulations, here we used hierarchical architecture for routing in order to route packets between wireless and wired domains. The routing information for wired nodes is based on connectivity of the topology. This connectivity information between the nodes is used to generate the forwarding tables in each wired node. Because of the dynamic properties, wireless nodes do not follow the concept of "links". In a wireless topology, packets are routed using their ad hoc routing protocols which can generate a forwarding tables by exchanging routing queries among its neighbors. So in this paper, we use base-stations as gateways between the two domains in order to exchange packets among these wired and wireless nodes. We set up hierarchical topology structure based on different domain and clusters. Next we setup tracing for the simulation for both wired and wireless domains as described in [20].

Next, we created the wired, wireless and base-station nodes. We set up the simulation environment by the help of [20]. Base station should have a different wired routing mechanism for configuring the Routing ON and OFF as it will act as gateway between wired and wireless domains. All other node config options used for the base-station remain the same for mobile node. Also, the BS(0) node is assigned as the base-station node for all the mobile nodes in the wireless domain, so that all packets originating from mobile nodes and destined for outside the wireless domain, will be forwarded by mobile nodes to their assigned base-station. Note that it is important for the base-station node to be in the same domain as the wireless nodes. According to [44], all packets originating from the wired domain, and destined for a wireless node, will reach the base-station which then uses its ad hoc routing protocol to route the packet to its correct destination. Thus, in a mixed simulation involving wired and wireless nodes, it is necessary to:

1. turn on hierarchical routing [20].
2. create separate domains for wired and wireless nodes; there may be multiple wired and wireless domains to simulate multiple networks [20].
3. have one base-station node in every wireless domain, through which the wireless nodes may communicate with nodes outside their domain [20].

Note that traffic flow for mobile nodes is not as yet supported in nam [20]. In trace file, we see traces for both wired domain and wireless domain (preceded by "WL" for wireless). Also note that the node-ids are created internally by the simulator and are assigned in the order of node creation. Actually, here we set up the simulation environment by the help of NS-manual [20].

### 6.2. Running mobile IP in the wired-cum-wireless topology

We run our simulation by the help of [20]. In the first scenario, we have created a wired-cum-wireless topology and have exchanged packets between a wired and wireless domain via a base station. But a mobile node may roam outside the domain of its base-station and should still continue to receive packets destined for it. In other words, it is necessary to extend mobile IP support in this wired-cum-wireless scenario. For this Mobile IP scenario, we have the same wired domain consisting of 2 wired nodes, W0 and W1. In addition we have 2 base-station nodes named Home Agent (HA) and Foreign Agent (FA) respectively. The wired node W1 is connected to HA and FA as shown in the figure below. There is a

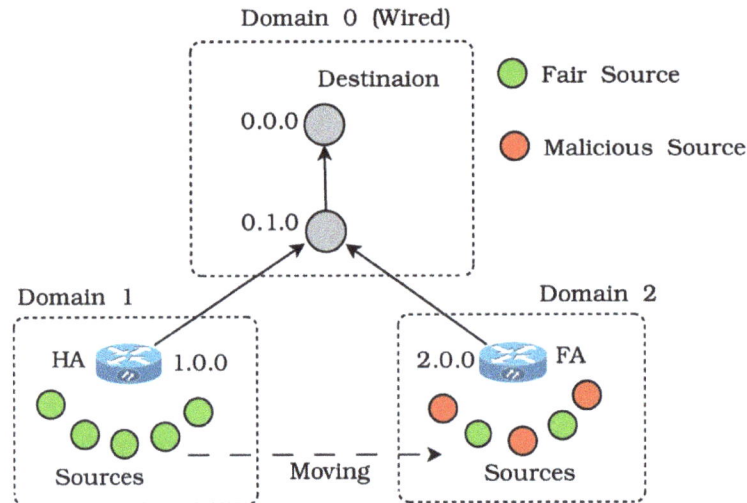

Fig. 4. Simulation of Mobile IP with DoS attack.

roaming mobile node called Mobile Node (MN) that moves between its home agent and foreign agents. We set up a TCP flow between W0 and MN. As MN moves out from the domain of it's HA, into the domain of FA, the packets destined for MN is redirected by its HA to the FA as per mobile IP protocol definitions. In the Fig. 4 the topology described above is shown.

In this topology, we have one wired domain (denoted by 0) and 2 wireless domains (denoted by 1 and 2 respectively). The wired node addresses remain the same, 0.0.0 and 0.1.0. In the first wireless domain (domain 1) we have base-station, HA and mobile node MN, in the same single cluster. Their addresses are 1.0.0 and 1.0.1 respectively. For the second wireless domain (domain 2), we have a base-station FA with an address of 2.0.0. However, in the simulation, the MN will move into the domain of FA and the packets originating from a wired domain and destined for MN will reach it as a result of the Mobile IP protocol. Wired nodes will be created as done earlier. However, in place of a single base station node, a HA and FA will be created. Note here that in order to turn on the mobile IP flag, we have configured the node structure accordingly using option mobile IP ON.

Next, we have created the Mobile Node (MN) as follows. We have to turn off the option wired Routing (used for creation of base-station nodes) before creating mobile nodes. Also, the HA is set up as the home-agent for the Mobile Node. The MN has an address called the care-of-address (COA). Based on the registration/beacons exchanged between the MN and the base-station node (of the domain the MN is currently in), the base-station's address is assigned as the MN's COA. Thus in this simulation, the address of the HA is assigned initially as the COA of MN. As MN moves into the domain of FA, its COA changes to that of the FA. We can see from the nam output that, initially the TCP packets are handed down to MN directly by its HA. As MN moves away from HA domain into the domain of the FA, we find the packets destined for MN, being encapsulated and forwarded to the FA which then decapsulates the packet and hands it over to the MN.

### 6.3. Implementation of mobile IP communication with denial of service attack

#### Scenario-1: Simulation of Mobile IP with DoS attack

In this scenario as shown in Fig. 4, there is a wired domain consisting of two wired nodes in two clusters with hierarchical addresses 0.0.0 and 0.1.0 respectively. Home Agent (HA) and Foreign Agent

Table 2
Simulation parameters for scenario 1

| Parameters | Values/Ranges |
|---|---|
| Speed of mobile nodes | 20 m/s |
| Packet size | 1000 Bytes |
| Transport agent | TCP |
| Application | FTP |
| Number of nodes (max) | 100 |
| Number of domains | 4–5 |
| Number of clusters | 5–10 |
| Simulation time | 60s |

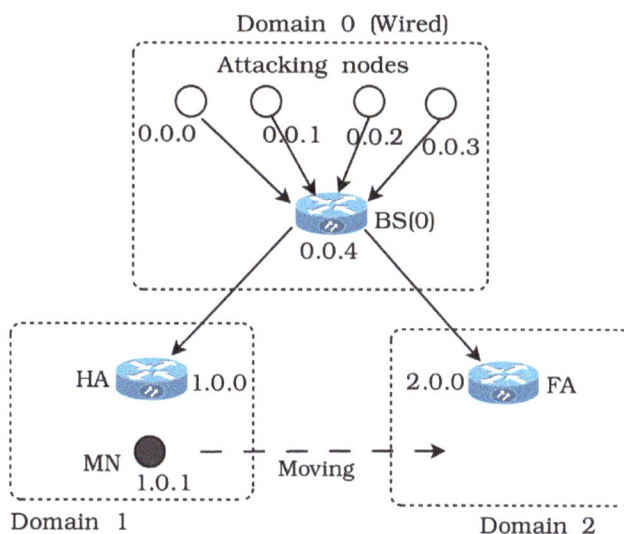

Fig. 5. Simulation of Mobile IP with DDoS attack.

(FA) are two base-station nodes in two other different domains. There are five roaming mobile nodes marked as green, which move between their home agent and foreign agent. There are five TCP flows from each of these five nodes where a node in the wired domain of address 0.0.0 is the destination for all. As the five sources move out from the domain, the HA to the domain of FA, the packets destined for the wired node are redirected by its HA to the FA as per mobile IP protocol definitions.

When the five sources reach the FA, they first get registered. In our simulation scenario, some of the sources become malicious (marked as red) and use spoofed addresses to attack the wired node of address 0.0.0.

The parameters used to implement this scenario are given in Table 2 below:

*Scenario-2: Simulation of Mobile IP with DDoS attack*

For the second attack scenario (DDoS attack) we have modified Domain 0 of the topology described above slightly. The modified topology is shown in Fig. 5.

Figure 5, wired nodes connected to BS (0) simultaneously send packets to the Mobile Node (MN) of domain 1. All these packets go to the MN through the base-station (HA) of domain 1. All these packets simultaneously overflow the queue of HA. Hence, the 4 nodes act as malicious nodes and try to consume network resources of the base station node (HA). So they are interrupting normal communications

Table 3
Simulation parameters for scenario 2

| Parameters | Values/Ranges |
|---|---|
| Packet size | 40Bytes |
| Transport agent | TCP |
| Application | CBR |
| Number of domains | 3 |
| Number of clusters | 4 |
| Simulation time | 60s |

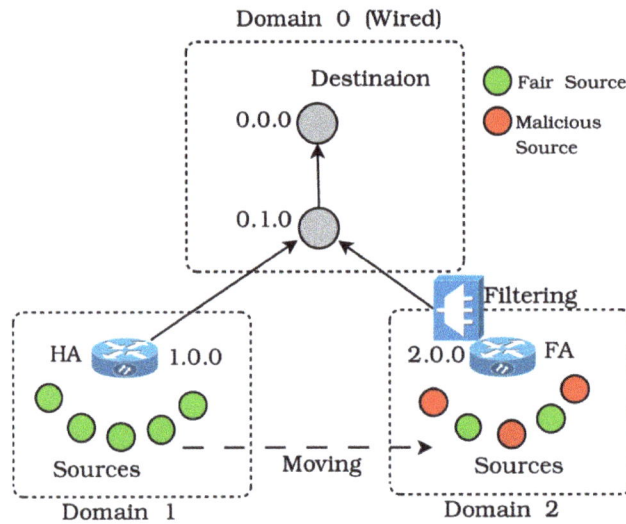

Fig. 6. Simulation with filtering in the domain periphery router.

through the base station node (HA). When fair nodes try to communicate with the base-station node HA or with the mobile node (MN), they cannot access services because of the attack on the base station

The parameters used for the simulation of this scenario are given in Table 3.

### 6.4. Implementation of Mobile IP communication with proposed technique

To protect against Denial of Service attack we have implemented a packet filtering technique in two vulnerable positions of the Mobile IP communication network, first in the domain periphery router and then in the base station node of the mobile node which is the receiver of the packets.

#### 6.4.1. Filtering in the Domain periphery router

According to our proposed solution, we cache the addresses of each node at base-station (FA) and when they start using spoofed IP addresses, those packets from unknown sources to the base-station (FA) are dropped as shown in Fig. 6. Although this mechanism consumes some memory of the base station, it can significantly reduce the chance of DoS attack at a very early stage.

The parameters used to implement this scenario are given in Table 4.

#### 6.4.2. Filtering in Base-station node

In this proposal we consider the case, if the attacker resides inside the same domain of the victim, then the edge or periphery router could not detect the attacking packet. That's why we have proposed

Table 4
Simulation parameters for filtering in domain periphery router

| Parameters | Values/Ranges |
| --- | --- |
| Speed of mobile nodes | 20 m/s |
| Packet size | 1000 Bytes |
| Transport agent | TCP |
| Application | FTP |
| Number of nodes (max) | 100 |
| Number of domains | 4–5 |
| Number of clusters | 5–10 |
| Simulation Time | 60s |

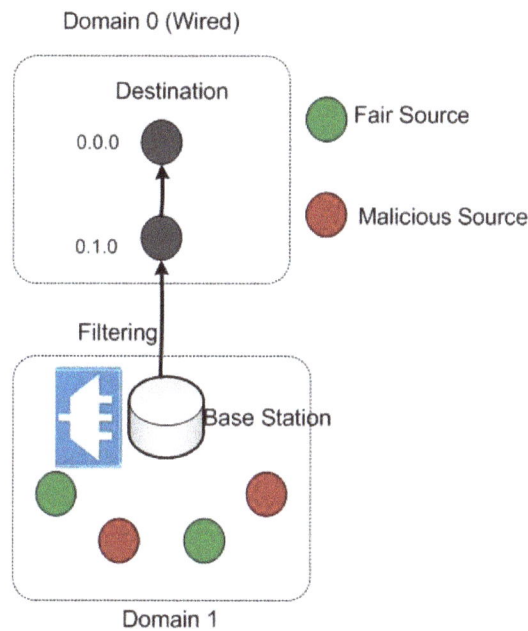

Fig. 7. Filtering in the base station node.

an additional filtering technique in the Base Station node (HA or FA) to which the mobile nodes are connected. Basically the base station nodes (HA or FA) in mobile IP communication are the main targets of the attackers, because the mobile nodes get services from these base stations. So detecting and preventing attacks in the base station nodes is very important. In our proposed scheme the base station node will filter a packet which will pass through the base station in the network as shown in Fig. 7.

The parameters used for the simulation of this scenario are given in Table 5 below:

### 6.4.3. Filtering in the base-station node by monitoring the queue

As stated earlier, filtering in the queue of an appropriate node can significantly reduce the effect of DoS attack. For example, in the case of a TCP SYN flooding attack, many attacking nodes start from the three way handshaking phase, but do not complete the phase. The size of the packets in the handshaking phase is very small. However, when many malicious nodes start a TCP SYN flooding attack on a mobile node, the base-station of that node observes a sudden increase of TCP SYN packets in its queue.

Table 5
Simulation parameters for filtering in base station node

| Parameters | Values/Ranges |
|---|---|
| Simulation area | $900 \times 900$ m |
| Speed (m/s) | 1 m/s to 20 m/s |
| Packet rate | 5 packets/ s |
| Packet size | 128 Bytes |
| Traffic source | CBR |
| Pause time | 60s |
| Routing protocols | DSDV and Mobile Ip |
| Number of nodes | 80–100 |
| Number of domains | 2–3 |
| Number of clusters | 3–4 |
| Transmission range | 400 m |
| Simulation time | 250 s |

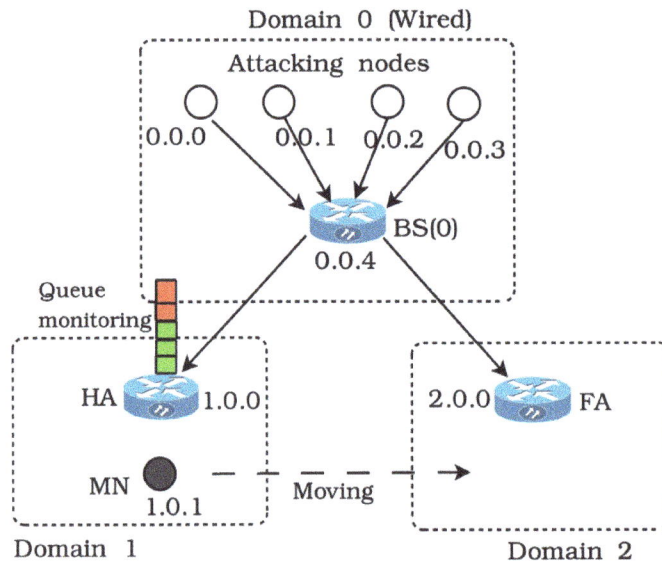

Fig. 8. Monitoring queue in the base station node.

In this scenario, four malicious nodes begin simultaneously attacking a mobile node under the base-station which is its home agent.

Figure 8 shows the simulation scenario of monitoring the queue in the mobile host's base station. According to our proposition, we filter the queue in the base-station by limiting the queue size to four and observe the queue status. In this simulation, we could drop 63 TCP SYN packets and thus the effect of the attack is reduced at the base-station node. It is obvious that this type of queue filtering can also be applied to the foreign agent. Hence, it is also possible to prevent the attack while the target node is moving from his home agent to the foreign agent.

The parameters used for the simulation of this scenario are given in Table 6 below:

Here we have used constant bit rate (CBR) applications to simulate a TCP SYN flooding attack where packet size is 40 bytes. In ns2, there are many parameters to express the status of a queue such as queue size in bytes, queue size in number of packets, the entrance and departure of packets etc. Here, we measured the queue of the base-station by number of packets.

Table 6
Simulation parameters for queue monitoring in the base station node

| Parameters | Values/Ranges |
|---|---|
| Packet Size | 40Bytes |
| Transport agent | TCP |
| Application | CBR |
| Number of Domains | 3 |
| Number of Clusters | 4 |
| Simulation Time | 60s |

No. of packet received (at every 5 seconds) with
and without filtering

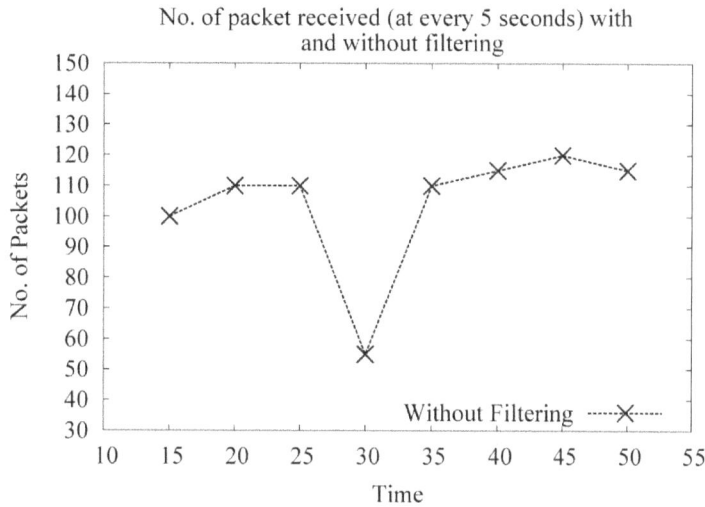

Fig. 9. Without applying the filtering technique.

## 6.5. Experimental result analysis

We make the following assumptions for the simplicity of simulation:

### 6.5.1. Comparison for filtering in the domain periphery router:

Figure 9 shows the resulting graph after simulating the Mobile IP communication without applying our proposed filtering technique in the domain periphery router. Implementation of this scenario is described in scenario-1 of section 3.1. From this graph we see the number of packets leaving the periphery router (FA) of domain 2. In this figure, there is a great fall in the curve at the time 25–35. When the Mobile Node (MN) moves away from its home agent, it loses its previous registration with the home agent. Before registering with the foreign agent, for an amount of time it deserves no registration at all. At this time, all the packets destined for the mobile node are dropped. For this reason, there is a great decline in the graph.

Figure 10 shows the resulting graph after simulating the Mobile IP communication by applying our proposed filtering technique in the domain periphery router. Implementation of this scenario is described in step 1 of section 5.3. We applied packet filtering in the base-station (FA) of the domain. The mobile nodes come to the foreign network and are registered with the base-station (FA). After that, some malicious packets are dropped by the periphery router (FA) of domain 2. From the graph, we can see that after registration with a foreign agent, the number of outgoing packets is less compared to that of the previous graph.

No. of packet received (at every 5 seconds) with
and without filtering

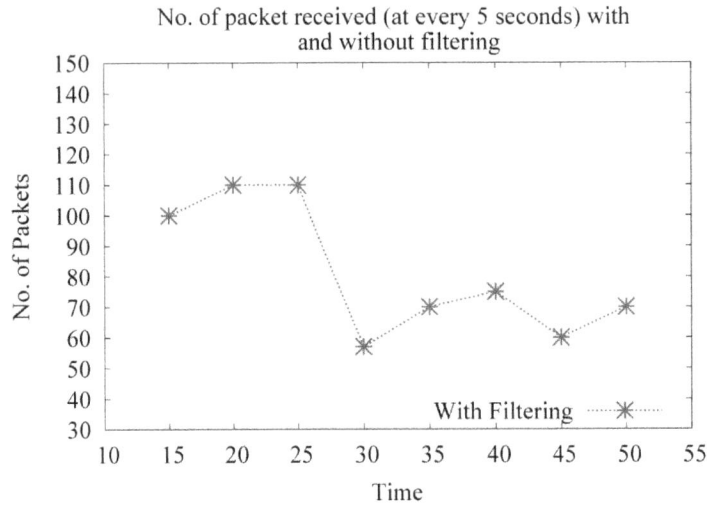

Fig. 10. With applying the filtering technique.

No. of packet received (at every 5 seconds) with
and without filtering

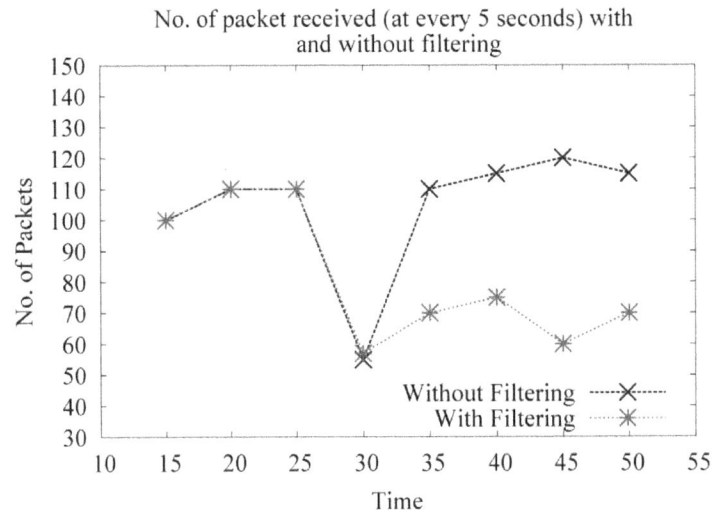

Fig. 11. Comparison of packet passing, with and without filtering technique applied.

A comparison of the above two graphs, that is with and without applying the filtering technique, is shown in Fig. 11.

Here we observe that, after the filtering technique was applied, the number of packets passed is reduced significantly. This is because the packets sent by the nodes with spoofed addresses are identified and dropped. Here it is noticeable that, using this filtering technique, packets will be dropped proportionally with the increase or decrease in the number of attacking nodes.

### 6.5.2. Comparison for filtering in the base station node

In our simulation the home agent router of the domain 1 is considered as the base station node and the mobile node connected to this base station is under attack.

Figure 12 shows that our system will exhibit better performance for malicious node detection rate if

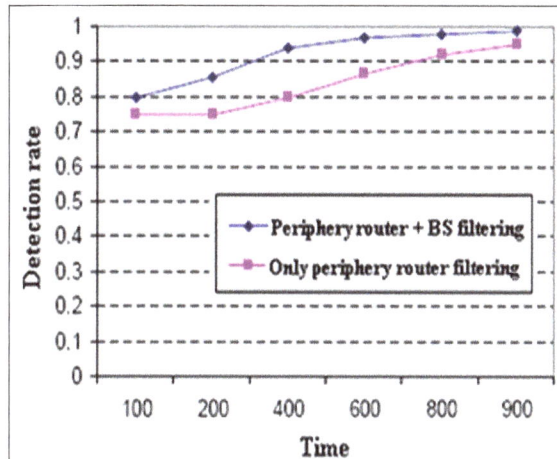

Fig. 12. Time Vs detection rate.

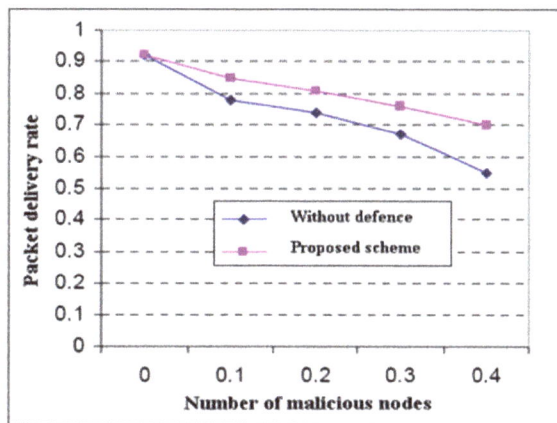

Fig. 13. Packet delivery Vs malicious nodes.

we use base station filtering and the periphery router filtering rather than using only periphery router filtering.

If number of misbehaving nodes increases then the packet delivery ratio will decrease due to attack in the servers and network resources consumed by the attackers. Figure 13 shows that if our proposed scheme is applied then the packet delivery ratio will increase slightly in spite of the presence of DoS attack.

Figure 14 shows that, as the network size increases the total overhead increases. When our proposed scheme is applied the overhead is relatively lower due to the use of clustering architecture.

Figure 15 shows the comparison between our proposed scheme and scheme without defense. Here the output shows the detection rate, packet delivery rate and overhead. This figure shows the summary of our proposed scheme. From this figure, we get that if we apply both periphery router and base station filtering, the detection rate increases than the only periphery filtering scheme. Our proposed scheme shows better packet delivery against malicious nodes and low communication overhead which is desirable.

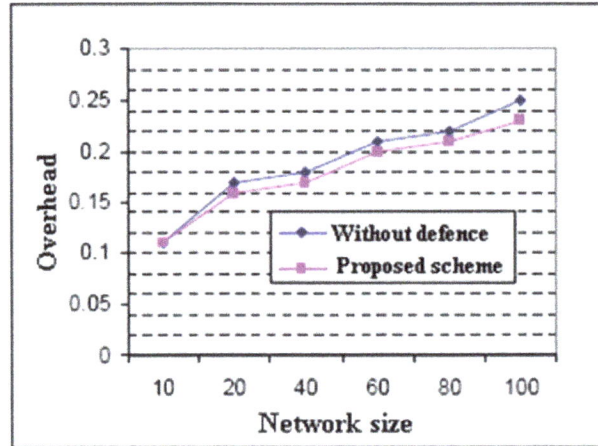

Fig. 14. Network size Vs overhead.

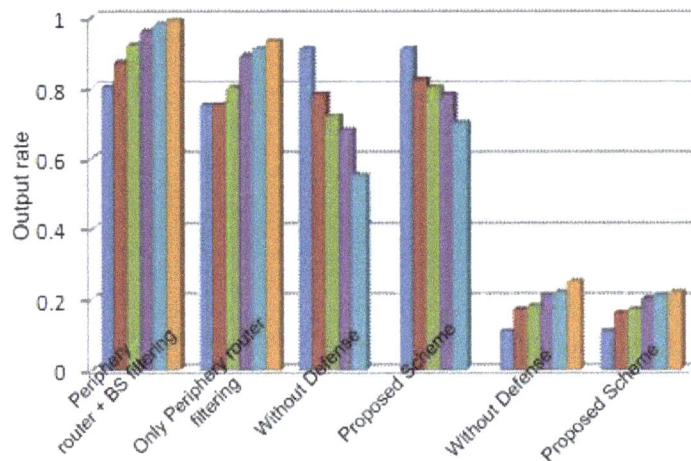

Fig. 15. Without defense vs. proposed scheme.

### 6.5.3. Comparison for queue monitoring in the base station node

In this section we show that queue status of base station while being monitored and while being attacked.

Figure 16, we observe that, some high fluctuations in queue size (measured in terms of number of packets). The peak values indicate the sudden presence of huge number of small packets which is a sign of DDoS attack. The peak values rose to almost 15 in this figure.

Figure 17 is the resulting graph after applying our proposed queue monitoring technique in the base station node. In this simulation the queue of the home agent router of domain 1 is monitored because it is attacked as a victim by the attackers.

Here we observe the queue status of the base-station node while it is being filtered by limiting the queue size to an adaptive value; in our case it was set to 4. Limiting the queue size results in some packets drops which were used for DDoS attack here. In this simulation, we found by analyzing the trace file of the simulation, that 63 malicious packets are dropped after using this filtering technique.

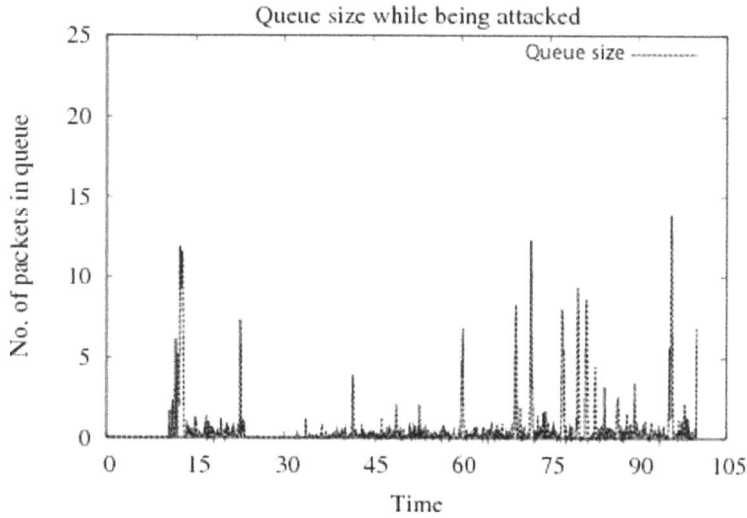

Fig. 16. Queue status of base station while being attacked.

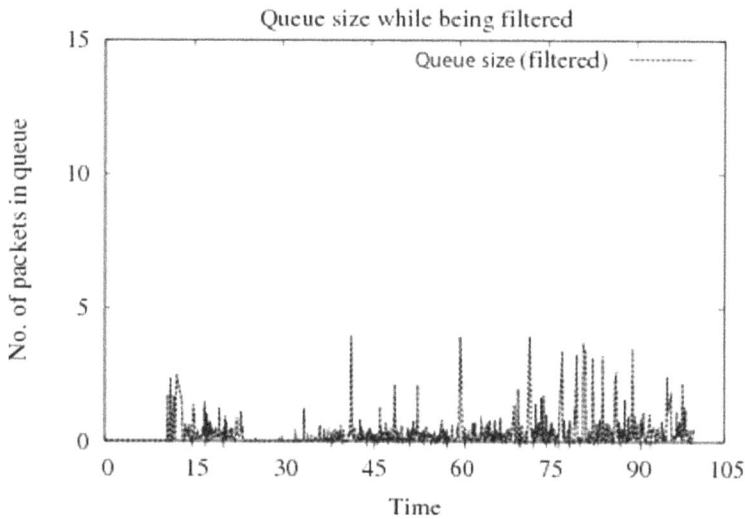

Fig. 17. Queue status at base station while being filtered.

### 6.5.4. Comparison with other schemes

In this section, we compare our proposed schemes with others. Analysis and comparisons are made among the approaches that belong to an identical technique category which ensures security by mitigating DoS attack in Mobile IP communication. Zao et al. [50] proposed a public-key based security schemes for mobile IP but this scheme increases communication overhead like all other conventional PKI techniques whereas we showed our proposed scheme is low in communication overhead. Park et al. [36] proposed route-based packet filtering technique for preventing DDoS attack but this approach is suitable only for internet based system whereas our scheme is suitable for all mobile IP systems. Ferguson et al. [17] proposed ingress filtering technique but this approach only works for Internet Service Providers and for internet based community whereas our approach works for all mobile IP based communication. Like other scheme, our proposed scheme has some shortcomings as our scheme only works for cluster-based

Table 7
Comparison between our scheme and other schemes

| Proposed schemes | Techniques | Shortcomings |
|---|---|---|
| Public key based secure mobile Ip [50] | Public key management techniques | Like all other conventional PKI techniques, it increases communication overhead |
| Route based filtering approach [36] | Route based distributed packet filtering technique | This approach is only suitable for internet based system |
| Ingress filtering to [10] mitigate DoS attack | Ingress filtering techniques | This approach works for Internet Service Providers and for internet based community |
| Our proposed schemes | Filtering and queue monitoring techniques at base station | It only works for clustering architecture |

architecture. A comprehensive comparison between our scheme and other's schemes are shown in Table 7.

## 7. Conclusion and future works

Denial of Service attack in mobile IP communication is considered to be one of the most severe attacks. The detection and prevention of this attack is more difficult in cases of mobile IP communications than in their wired counterparts. In this paper, we proposed a technique for detecting and preventing DoS attacks in mobile IP communication. We propose to use packet filtering techniques that work in different domains and base stations of mobile IP communication to detect suspicious packets and to improve the performance. If any packet contains a spoofed IP address which is created by DoS attackers, our scheme can detect this and then filter the suspected packet. The proposed system can mitigate the effect of Denial of Service (DoS) attack by applying three methods which are elaborately described in this work: (i) by filtering in the domain periphery router (ii) by filtering in the base station and (iii) by queue monitoring at the vulnerable points of base-station node. We proposed to apply a packet filtering technique at the vulnerable points of the Mobile IP network. We used the network simulator NS-2 for simulating and evaluating the performance of our proposed system. We observed that the performance of our proposed system is better than the system without any protection. We also observed that our proposed technique can significantly reduce the effect of DoS and DDoS attacks.

## References

[1]   C. Barros, A proposal for ICMP traceback messages. Internet Draft http://www.research.att.com/lists/ietfitrace/2000/09/msg00044.html, Sept. 18, 2000.
[2]   S.M. Bellovin, M. Leech and T. Taylor, ICMP traceback messages. Obsolete Internet draft, February 2003.
[3]   H. Burch and B. Cheswick, Tracing anonymous packets to their approximate source, *Proceedings of the 14th USENIX conference on System administration*, 2000.
[4]   T. Braun and M. Danzeisen, Secure mobile IP communication, *In Conference on Local Computer Networks* (2001), 586–593.
[5]   T. Braun and M. Danzeisen, Access to Mobile IP Users to Firewall Protected VPNs, *Proceedings of Workshop on Wireless Local Networks at the 26th Annual IEEE Conference on Local Computer Networks*, 2001.
[6]   C. Brenton, What is Egress Filtering and How can I Implement it?. Published by the SANS Institute, June 13, 2011.
[7]   H.B. Chang, H.J. Kwon and J.G. Kang, The design and implementation of tamper resistance for mobile game service, *Mobile Information System* 6(1) (2010), 85–105.
[8]   C.L. Chen, Design of a secure RFID authentication scheme preceding market transactions, *Mobile Information Systems* 7(3) (2011), 201–216.

[9]   Y.F. Ciou, F.Y. Leu, Y.L. Huang and K. Yim, A handover security mechanism employing the Diffie-Hellman key exchange approach for the IEEE802.16e wireless networks, *Mobile Information Systems* **7**(3) (2011), 241–269.

[10]  L. Dang, W. Kou, H. Li, J.Z hang, X. Cao, B. Zhao and K. Fan, Efficient ID-based registration protocol featured with user anonymity in mobile IP networks, *IEEE Transactions on Wireless Communications* **9**(2) (2010), 594–604.

[11]  T. Delot, S. Ilarri, N. Cenerario and T. Hien, Event Sharing in Vehicular Networks Using Geographic Vectors and Maps, *Mobile Information Systems* **7**(1) (2011), 21–44.

[12]  M.K. Denko, Detection and Prevention of Denial of Service (DoS) Attacks in Mobile Ad Hoc Networks using Reputation-Based Incentive Scheme, *Journal of Systemics, Cybernetics and Informatics* **3**(4) (2005), 1–9.

[13]  R.H. Deng, J. Zhou and F. Bao, Defending against redirect attacks in mobile IP, *Proceedings of the 9th ACM Conference on Computer and Communications Security* (2002), 59–67.

[14]  I. Doh, J. Lim and K. Chae, Distributed authentication mechanism for secure channel establishment in ubiquitous medical sensor networks, *Mobile Information Systems* **7**(3) (2011), 189–200.

[15]  A. Durresi, M. Durresi and B. Barolli, Secure authentication in heterogeneous wireless networks, *Journal of Mobile Information Systems* **4**(2) (2008), 119–130.

[16]  K. Fall and K. Vardhan, ns notes and documentation, available from http://www-mash.cs.berkeley.edu/ns/, 1999.

[17]  P. Ferguson and D. Senie, Network Ingress Filtering: Defeating Denial of Service Attacks which employ IP Source Address Spoofing, RFC 2827, BCP 38, 2000.

[18]  H.L. Flanagan and G.P.A. Version, Egress filtering-keeping the Internet safe from your systems. April 30, 2001. Available at SANS Institute. http://rr.sans.org/sysadmin/egress.php.

[19]  P. Fulop, S. Imre, S. Szabo and T. Szalka, Accurate mobility modeling and location prediction based on pattern analysis of handover series in mobile networks, *Journal of Mobile Information Systems* **5**(3) (2009), 255–289.

[20]  M. Greis, Tutorial for the Network Simulator NS, available online (11.08.2004), http://www.isi.edu/nsnam/ns/tutorial.

[21]  V. Gupta, S. Krishnamurthy and M. Faloutsos, Denial of service attacks at the MAC layer in wireless ad hoc networks, *Proceedings of IEEE MILCOM Conference* (2002), 1118–1123.

[22]  V. Gupta and G. Montenegro, Secure and mobile networking, *Journal of Mobile Networks and Applications – Special issue: mobile networking in the Internet* **3**(4) (1998), 381–390.

[23]  A. Habib, M. Hefeeda and B. Bhargava, Detecting service violations and DoS attacks, *Proceedings of Internet Society Symposium on Network and Distributed System Security*, 2003.

[24]  A.A. Hamidian, A study of internet connectivity for mobile ad hoc networks in ns-2, *Published by Department of Communication Systems, Lund Institute of Technology, Lund University*, 2003.

[25]  L.T. Heberlein and M. Bishop, Attack class: Address spoofing, *Proceedings of the 19th National Information Systems Security Conference* (1996), 371–377.

[26]  A. Inoue, M. Ishiyama, A. Fukumoto and T. Okamoto, Secure mobile IP using IP security primitives, *Proceedings Sixth IEEE workshops on Enabling Technologies: Infrastructure for Collaborative Enterprises* (1997), 235–241.

[27]  R. Islam, Enhanced security in Mobile IP communication, *PhD thesis, Stockholm University*, 2005.

[28]  C. Jin, H. Wang and K.G. Shin, Hop-count filtering: an effective defense against spoofed DDoS traffic, *Proceedings of the 10th ACM Conference on Computer and Communications Security* (2003), 30–41.

[29]  H. Kim and J.H. Lee, Diffie-Hellman key based authentication in proxy mobile IPv6, *Journal of Mobile Information Systems* **6**(1) (2010), 107–121.

[30]  P. Lin, S.M. Cheng and W. Liao, Modeling Key Caching for Mobile IP Authentication, Authorization, and Accounting (AAA) Services, *IEEE Transactions on Vehicular Technology* **58**(7) (2009), 3596–3608.

[31]  L. Liu, A Client-Transparent Approach to Defend Against Denial of Service Attacks, *Proceedings of 25th IEEE Symposium on Reliable Distributed Systems* (SRDS 06), 2006.

[32]  G. Montenegro, Reverse tunneling for Mobile IP, revised. 2001.

[33]  P. Nikander, J. Arkko, T. Aura, G. Montenegro and E. Nordmark, Mobile IP version 6 (MIPv6) route optimization security design, *Proceedings OF THE IEEE Vehicular Technology Conference* (2003), 2004–2008.

[34]  P. Owezarski, On the impact of DoS attacks on Internet traffic characteristics and QoS, 14th International Conference on Computer Communications and Networks, ICCCN 2005, 17–19 Oct. 2005, pp. 269–274.

[35]  F. Palmieri, U. Fiore and A. Castiglione, Automatic security assessment for next generation wireless mobile networks, *Mobile Information Systems* **7**(3) (2011), 217–239.

[36]  K. Park and H. Lee, A proactive approach to distributed DoS attack prevention using route-based packet filtering, *Proceedings of ACM SIGCOMM Conference*, 2001.

[37]  S. Parvin, S. Ali, S. Han and T. Dillon, Security against DOS attack in mobile IP communication, *Proceedings of the 2nd International Conference on Security of Information and Networks* (2010), 152–157.

[38]  S. Parvin, S. Ali, J. Singh, H. Hussain and S. Han, Towards DoS Attack Prevention based on Clustering Architecture in Mobile IP Communication, *Proceedings of IEEE IECON* (2009), 3183–3188.

[39]  C.E. Perkins and D.B. Johnson, Route optimization for mobile IP, *Cluster Computing* **1**(2) (1998), 161–176.

[40]  C.E. Perkins, Mobile IP: Design Principles and Practices, *Addison-Wesley*, 1997.

[41]  N.N. Qadri, M. Altaf, M. Fleury and M. Ghanbari, Robust video communication over an urban VANET, *Mobile Information Systems* **6**(3) (2010), 259–280.

[42]  B.M. Reshmi and S.S. Manvi, Bhagyavati. An agent based intrusion detection model for mobile ad hoc networks, *Journal of Mobile Information Systems* **2**(4) (2006), 169–191.

[43]  Security aspects of Mobile IP. SANS Institute 2001, as part of the information security reading room.

[44]  D. Strom and S.R. Room, The Packet Filter: A Basic Network Security Tool. September 2000, Available at http://www.giac.org/paper/gsec/131/packet-filter-basic-network-security-tool/100197.

[45]  G. Tuquerres, M.R. Salvador and R. Sprenkels, Mobile IP: security and application, *Telematics Systems and Services*, 1999.

[46]  C.H. Wu, A.T. Cheng, S.T. Lee, J.M. Ho and D.T. Lee, Bi-directional route optimization in mobile IP over wireless LAN, *Proceedings of IEEE 56th Vehicular Technology Conference*, 2002.

[47]  Y. Xiang and W. Zhou, A defense system against DDOS attacks by largescale IP traceback, *Proceedings of third International Conference on Information Technology and Applications*, 2005.

[48]  X. Yang, D. Wetherall and T. Anderson, A DoS-limiting network architecture, *Proceedings of the 2005 conference on Applications, Technologies, Architectures, and Protocols for Computer Communications* **35**(4) (2005), 241–252.

[49]  I. You, J.H. Lee, Y. Hori and K. Sakurai, Enhancing MISP with Fast Mobile IPv6 Security, *Mobile Information Systems* **7**(3) (2011), 271–283.

[50]  J. Zao, S. Kent, J. Gahm, G. Troxel, M. Condell, P. Helinek, N. Yuan and I. Castineyra, A public-key based secure Mobile IP, *Journal of Wireless Networks* **5**(5) (1999), 373–390.

[51]  J.K. Zao and M. Condell, Use of IPSec in mobile IP. 1997, 381–390.

---

**Sazia Parvin** received her MS degree in Computer Engineering from Korea Aerospace University in 2008. Presently she is a PhD student at Digital Ecosystems and Business Intelligence Institute, Curtin University, Australia. Her research interests include security in wireless communications and networking. She is a lecturer at Department of Computer Science and Engineering in Dhaka University, Dhaka, Bangladesh.

**Farookh Khadeer Hussain** received the Bachelor of Technology degree in computer science and computer engineering; the M.S. degree in Information Yechnology from the La Trobe University, Melbourne, Australia; and the Ph.D. degree in Information Systems from Curtin University of Technology, Perth, Australia, in 2006. He is currently a Research Fellow with the Digital Ecosystems and Business Intelligence Institute (DEBII), Curtin University, Perth, Australia. His areas of active research are trust, reputation, trust ontologies, data modeling of public and private trust data, semantic web technologies and industrial informatics. He works actively in the domain of making informed business decisions (business intelligence) through the use of trust and reputation technology. He is interested in the application of trust and reputation as a technology, as a business analysis and intelligence tool, and the applications of trust and reputation to various domains.

**Sohrab Ali** received his Bachelor and MS degree from Computer Science and Engineering department in Dhaka University. Presently he serving as a lecturer in People's University in Dhaka, Bangladesh. His research interests include security in wireless communications and networking.

# 2

# A lifelog browser for visualization and search of mobile everyday-life

Keum-Sung Hwang and Sung-Bae Cho*

*Department of Computer Science, Yonsei University, Seoul, Korea*

**Abstract.** Mobile devices can now handle a great deal of information thanks to the convergence of diverse functionalities. Mobile environments have already shown great potential in terms of providing customized service to users because they can record meaningful and private information continually for long periods of time. The research for understanding, searching and summarizing the everyday-life of human has received increasing attention in recent years due to the digital convergence. In this paper, we propose a mobile life browser, which visualizes and searches human's mobile life based on the contents and context of lifelog data. The mobile life browser is for searching the personal information effectively collected on his/her mobile device and for supporting the concept-based searching method by using concept networks and Bayesian networks. In the experiments, we collected the real mobile log data from three users for a month and visualized the mobile lives of the users with the mobile life browser developed. Some tests on searching tasks confirmed that the result using the proposed concept-based searching method is promising.

Keywords: Life browser, mobile lifelog, context-aware computing, concept-based search

## 1. Introduction

Growth of technology for devices such as smart phones and web has facilitated to expand the availability of ubiquitous devices to produce, store, view, and exchange lifelog that means the log collected around a human-life. Almost everyone is not only a consumer but also a producer of information in a world in which enormous amount of contextual information can be recorded automatically by the various devices we use. The information with human's life, which is called lifelog, can raise diverse intelligent and personal services.

The mobile device like cell phone is one of the most viable for the lifelog services. Mobile environments have very different characteristics from desktop computer environments. First of all, mobile devices can collect and manage various kinds of user information, for example, by logging user's calls, SMS (short message service), photography, music-playing and GPS (global positioning system) information. Also, mobile devices can be customized to fit any given user's preferences. Furthermore, mobile devices can collect everyday information effectively. Such features allow for the possibility of diverse and convenient services, so they have attracted the attention of researchers and practitioners. The research conducted by Nokia is a good example [1].

*Corresponding author: Sung-Bae Cho, Department of Computer Science, Yonsei University, Seoul, Korea.
E-mail: sbcho@cs.yonsei.ac.kr.

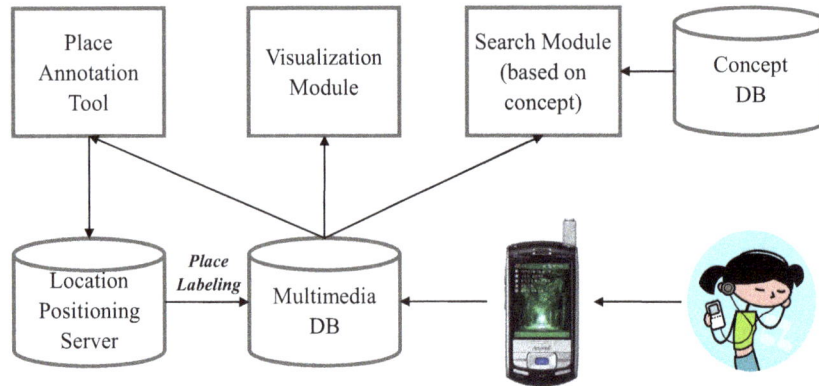

Fig. 1. General operations of mobile life browser.

In recent years, researchers have started making progress in effectively integrating context [2] and content for lifelog mining and management. Integration of content and context is crucial to human understanding of lifelog. Especially, the context-aware technique [3] that has been widely devised is more applicable to mobile environments, so many intelligent services such as intelligent calling services [4], messaging services [5], analysis, summary and visualization [6,7], collection and management of mobile logs [8–14] have been actively investigated.

In this paper, we investigated and developed a mobile life browser by making a database from the data collected on mobile devices and by analyzing and summarizing the log data and by visualizing the personal everyday-life efficiently. The mobile life browser enables us to manage personal lifelog data and supports to develop diverse additional services by utilizing the information easily.

Figure 1 shows the general operations of a mobile life browser. Since the logs gathered from mobile device are low level, we need to extract high-level symbolic information from them to develop diverse applications. We constructed a location positioning server (LPS) that transformed the GPS position coordinates (longitude, latitude) into symbolic location, and made the interface that enabled users to add semantic labels for certain position coordinates easily.

The mobile life browser supports basic search of log data, visualization and concept-based search capability by semantic networks. It converts the low-level log data into symbolic data using LPS, and provides relational information search capability using a common sense database (DB). The common sense DB is developed from the diverse common senses such as "library is for studying" from diverse internet users [15]. If a user inputs a concept query 'study' into the mobile life browser, he/she can get the contexts and places related to the concept. It supports the basis for easy development of concept inquiry, natural language question, automatic diary generation and automatic summarization function to convert low-level information into high-level using concept network and location positioning system. In this paper, we conducted usability tests for the mobile life browser based on real world data collected from three college students.

## 2. Backgrounds

### 2.1. Context-awareness

Context-awareness in computer science refers to the idea that computers can sense and react based on their circumstance and situation. Context is the information that devices may have about their environments [3]. Dey defined context and context awareness as "Context is any information that can be used to characterize the situation of an entity [2]. An entity is a person, place, or object that is considered relevant to the interaction between a user and an application, including the user and applications themselves" and "A system is context-aware if it uses context to provide relevant information and/or services to the user, where relevancy depends on the user's task [2]". Context aware systems are concerned with the acquisition of context, the abstraction and understanding of context, and application behavior based on the recognized context [16]. In this paper, we extract contexts from lifelog and utilize the context-awareness techniques for intelligent service for mobile users.

## 2.2. Lifelog

Numerous sources are available for logging information. For example, ContextPhone is context-logging software for Nokia 60 smart phones [17], and its source is available to the public. It collects information on a wide range of topics by logging photographs taken, music downloads, short-message-service use, multimedia messaging service use, and Bluetooth use; monitoring the phone's battery level; storing call logs; and recording applications in use.

The MIT Reality Mining group has developed a serendipity service, which cues informational, face-to-face interactions between nearby users who do not know each other but probably should. Their service uses the ContextPhone software [18], and they have been collaborating with the MIT Common Sense Reasoning group to generate diaries automatically. Because the research is still in the early stage, it has yet to produce any concrete results although the Reality Mining group has made available its visualization tool for a collected log. However, their work shows a new way of generating more interpretable high-level diaries using common sense.

MyLifeBits is a Microsoft research project to create a "lifetime store of everything". It includes full-text search, text and audio annotations, and hyperlinks. SenseCam is one of hardware developed for the project [13]. It stores photographs of everyday-life by monitoring the brightness and temperature of environment continuously.

LifeLog proposed by MIT reality mining group was a tool for visualizing the log information gathered by the logging software developed by Helsinki University. It was developed for simple automatic diary [21]. The Lifeblog software of Nokia afforded new opportunities to capture, manipulate and communicate daily events and thoughts that assisted in making meaning in individual and collective contexts [1].

However, because ContextPhone is based on the Symbian operating system (OS), we also developed a logging module for the Windows mobile environment. Our logging system runs on Windows Mobile OS with a small GPS receiver attached to the device. The system continuously records the user's latitude and longitude and lets users easily access call logs and the address book. It also stores SMS texts. We also modified the code for a photo viewer and an MP3 player and added it to our system to log usage information. We can easily gather a photo's creation time and other low-level information from the photo's header. The system retrieves weather information from the Korean Meteorological Administration (www.kma.go.kr), and it samples the GPS and battery level.

## 2.3. Life browser

Life browser is a tool for searching personal information effectively and is a term used for the first time by Horvitz of Microsoft Research [19]. Horvitz et al. proposed a method that detected and estimated

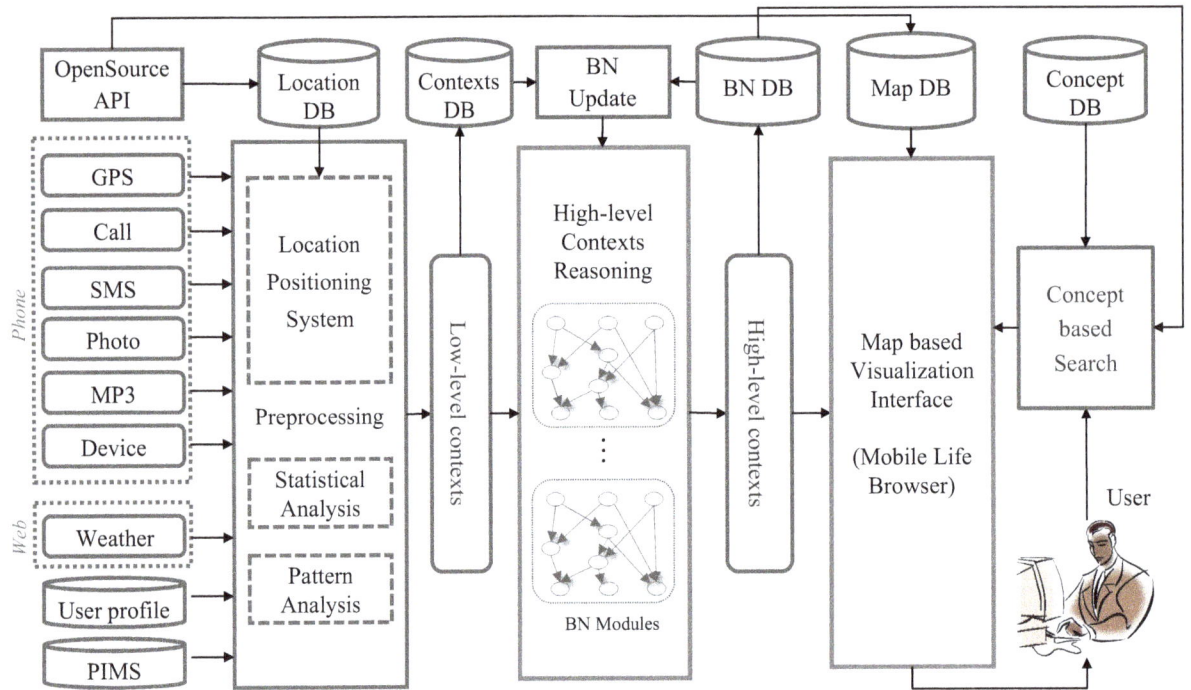

Fig. 2. The process of mobile life browser.

landmarks, which are used as a special event for helping recall it, by discovering a given human's cognitive activity model from PC log data based on the Bayesian approach [20]. Their approach visualized everyday-PC-life of humans with hierarchical structure and showed a good performance for recognizing and learning.

In [22], the authors proposed a method for identifying semantic information on mobile devices and developed AniDiary software making cartoon diary of mobile users automatically. The preliminary investigation is the foundation of the present work.

## 3. Mobile life browser

The overall process of mobile life browser is shown in Fig. 2. Various mobile lifelog data are preprocessed in advance, and then the context-reasoning module detects the high-level contexts from the generated low-level contexts. The preprocessing module operates with the techniques of statistical analysis, pattern recognition and rules. The context reasoning module works based on the probabilistic inference method of Bayesian networks [22]. The extracted context information is utilized in the application like a mobile life browser.

The map-based visualization module shows the map-based browser interface to the user and supports concept-based searching capability, since it is developed to work on web environment with using open-source API and database server.

### 3.1. Lifelog collection

A mobile device is easy to gather diverse information about the user since the user is always along with it. Table 1 shows the log information collected on a mobile device. The GPS log presents the

Table 1
The log information collected on a mobile device

| Log | Information |
| --- | --- |
| GPS | Latitude, longitude, velocity, direction, date, time |
| Call | Caller's phone number, call/receive/absence log, time span, start/end time |
| SMS | Sender's phone number, call/receive/absence log, time span, start/end time |
| Photographing | Photo file name, the time taken |
| Weather | Weather, visibility range (km), cloud degree (%), temperature ($^\circ$C), discomfort index, effective temperature ($^\circ$C), rainfall (mm), snowfall (cm), humidity (%), wind direction, wind velocity (m/s), barometer (hPa) |
| MP3 player | Title, time span, start/end time |
| Charging | Charging status, time span, start/end time |
| User profile | Gender, name, home address, phone number, and occupation of user |
| PIMS | Phone number of friends, colleagues and relatives |

places that the user visited, and the call and SMS logs show the user's calling patterns. The MP3 (a music file format) player log offers an idea of the user's emotions and the photograph log shows when the user wants to memorize something. Additionally, the user profiles and PIMS (personal information management system) datasets were used to find the user's social position (student or worker), gender, the position of their home, and the names and phone numbers of their friends and relatives.

### 3.2. Preprocessing

This stage employs standard statistical analysis to extract significant information. Because raw information is not meaningful, we use statistical variations to detect informative situations. To determine discrete information using SMS text, call logs, photos, and MP3 selections, we extract patterns using simple statistical techniques, such as calculating the average, maximum, and minimum values or the frequency over the time domain.

Since logs have temporal properties, we considered their time spans, frequencies (per hour, daily, or weekly), and start/end times as well as their motifs that reflect the density of given events. For example, the motif of time $t$, $M_t$, can be calculated as Eq. (1), where $x$ represents the event and $\alpha(x)$ and $\beta(x)$ represent the increment and decrement constants for event $x$, respectively.

$$M_t(x) = \begin{cases} M_{t-1}(x) + \alpha(x), & \text{if event } x \text{ is occurred} \\ M_{t-1}(x) - \beta(x), & \text{otherwise} \end{cases} \tag{1}$$

We infer a user's current position from the GPS value of a pair of longitude and latitude coordinates because a device can sometimes lose its GPS signal in a building or a shadowed area of an urban environment, or when placed in a pocket or bag. The coordinates from the GPS log are used to get place names. In this paper, we developed and utilized location positioning server to determine the user's current position. The GPS signal was cleared by the following preprocessing methods.

*Outlier clearing*: We used Eq. (2) to get rid of the peculiar GPS data from its normal boundary.

$$\text{IF} \left( \frac{d_{i-1,i}}{t_i - t_{i-1}} > k_v \text{ OR } \frac{d_{i,i+1}}{t_{i+1} - t_i} > k_v \right) \text{ THEN } Disregard(\vec{p}_i) \tag{2}$$

where $t_i$ is the time of the $i^{\text{th}}$ GPS data, $d_{a,b}$ means the distance between the $a^{\text{th}}$ and $b^{\text{th}}$ coordinates, and $p_i$ is the $i^{\text{th}}$ GPS coordinates. $k_v$ denotes the threshold for outlier clearing.

*Missing data correction*: The calculation of the missing GPS data is followed as Eq. (3).

$$\vec{p}_i = \vec{p}_{i-1} + \frac{t_{i+1} - t_i}{t_{i+1} - t_{i-1}} (\vec{p}_{i+1} - \vec{p}_{i-1}) \tag{3}$$

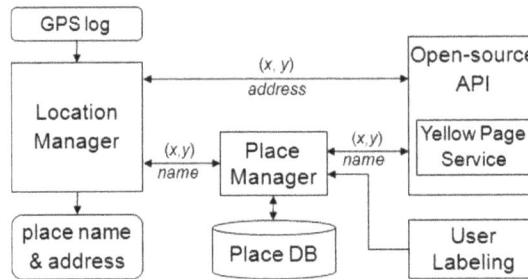

Fig. 3. The process of the location positioning system.

### 3.3. Location positioning system

Location information is useful since it may produce additional services. For example, there are map services of web portal such as Google Earth and Google local map service. They allow a user to explore the map of living area on the web, and to access satellite map images. The Google Maps also provides visualization service of commercial information around a certain building. Microsoft trades actually map information for European Union and North America [23].

Emerging proactive applications require to reason about "place", not coordinates [24–27]. The conventional systems rely on manually defining places which, while useful, are not practical. In this paper, we developed Location Positioning System (LPS), which converts the coordinates into symbolic location information automatically as shown in Fig. 3. Using the GPS value, a web service (an open-source API of Korea, http://maps.naver.com) identifies the nearest building. We then transform the raw information into semantic labels using stored information about the relationship between the GPS value and the semantic label (the building or street name). The stored information might include semantic labels such as "my home", "my office", or "my friend's home". The user can manually input the information using map-based graphic user interface for visualization.

The general location positioning method utilizes conversion based on the distance from a center point of an area, but this method is not fine-grained because it assumes that the area has a round shape expressed by the center point and diameter. To overcome the problem, the LPS database is constructed based on the unit of $1_{sec} \times 1_{sec}$ ($1_{sec}$ is 25 m∼30 m) block area in this paper.

### 3.4. High-level context reasoning based on Bayesian network

The high-level context reasoning module performs probabilistic inference with Bayesian network model. Since the problem of understanding human's intention and inference of contexts accompany much uncertainty, it is known that the probabilistic models such as Bayesian network have advantage for solving it [28]. In this paper, the Bayesian network model is also utilized for understanding the user's life and extracting meaningful information. For example, we can infer an activity context 'eating-out' from the evidences such as 'lunch-time' and a place information 'restaurant' with the Bayesian network shown in Fig. 4.

Bayesian networks refer to models that can express large probability distributions with relatively small costs to statistical mechanics. They have the structure of a directed acyclic graph (DAG) that represents the link (arc) relations of the node, and has conditional probability tables (CPTs) that are constrained by the DAG structure [25]. The Bayesian network in Fig. 4 shows a DAG structure, node name, state

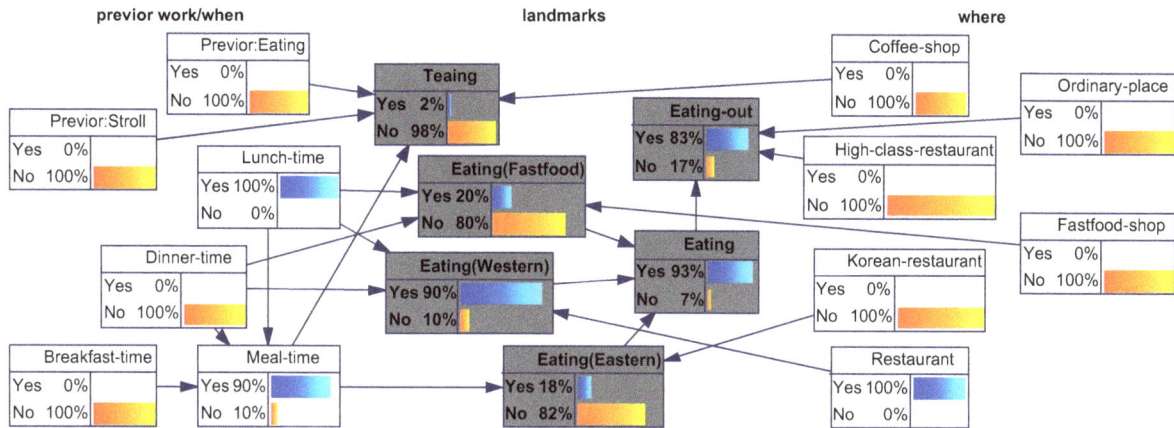

Fig. 4. The high-level context inference BN 'Activity in restaurant' designed and used in this paper. This is one of 39 BNs designed. The node of gray box is target high-level context node. This shows the case that the 'Lunch-time' and 'Restaurant' evidence are set.

name and inferred probabilities. On a Bayesian network, we can compute the belief value $Bel(h)$ for a hypothesis $h$ with given the evidence set $E$ by Bayes' rule as Eq. (4).

$$Bel(h) = P(h|E) = \frac{P(E|h)P(h)}{P(E)} = \frac{P(h \wedge E)}{P(E)} \tag{4}$$

The conditional probability value can be calculated by the Chain Rule as Eq. (5).

$$\begin{aligned}
P(x_1, x_2, \ldots, x_n) &= P(x_1, x_2, \ldots, x_{n-1})P(x_n|x_{n-1}, x_{n-2}, \ldots, x_1) \\
&= P(x_1)P(x_2|x_1)P(x_3|x_2, x_1)\ldots P(x_n|x_{n-1}, x_{n-2}, \ldots, x_1) \\
&= P(x_1)P(x_2|\pi_2)P(x_3|\pi_3)\ldots P(x_n|\pi_n)
\end{aligned} \tag{5}$$

where $x_i$ is the $i^{\text{th}}$ node and $\pi_i$ denotes the parent set of node $x_i$.

## 4. User interface of mobile life browser

The developed mobile life browser, called MyLifeBrowser, is a tool for visualizing and searching the lifelog data from mobile device. Figures 5 and 6 show the MyLifeBrowser working on the web browser. The MyLifeBrowser has four parts as follows.

1) *DB query panel*: This is on the right menu box and supports selection of current user and date. In the result of clicking the 'search' button, the trajectory mark is shown on the map area. The mark is yellow and represents the log per minute, so 1,440 marks can be expressed at a time.
2) *Contexts visualization*: The context checkbox at the left side of 'search' button determines whether it shows high-level contexts on map. The context covers each 30 minutes time span and it is shown in the map only if the reliability of the context is higher than a threshold.
3) *Filtering panel*: When a filter check box for a feature is checked, the color of the logs for the feature is changed. Figure 5 shows the case that the filter for 'mp3 listening' is checked. The type of the filter for weather and location is text since the type of them is also text.

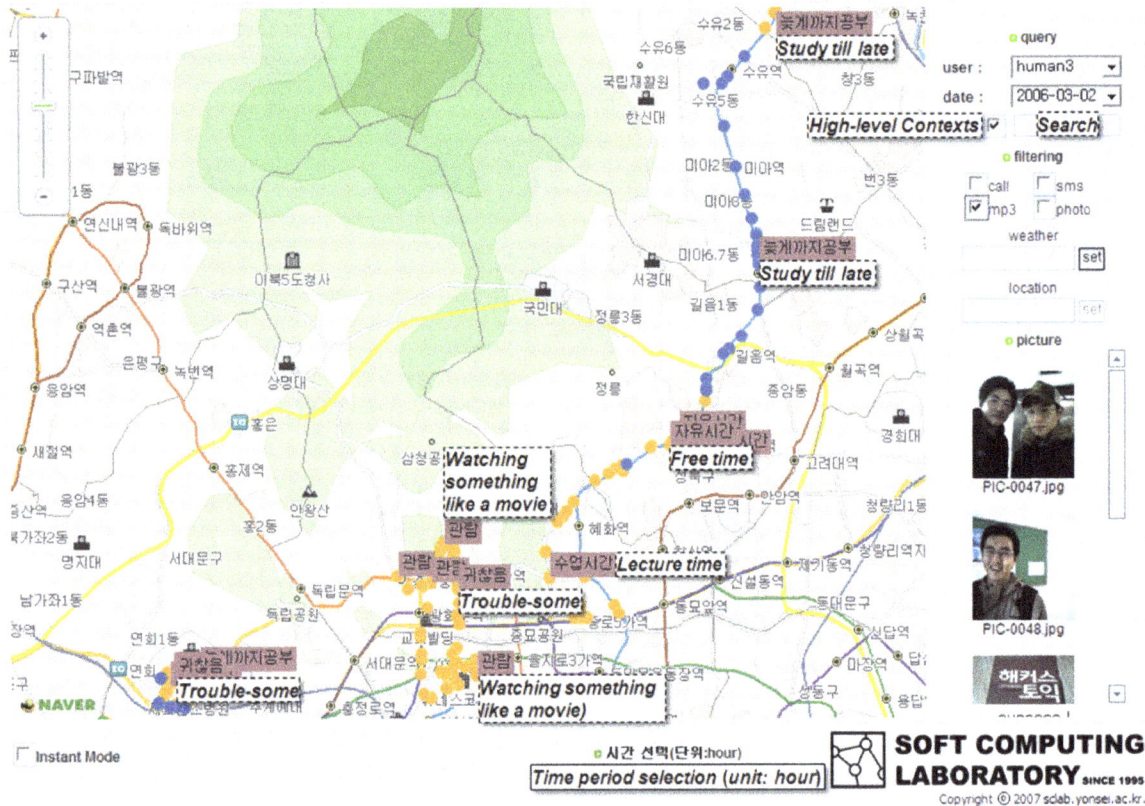

Fig. 5. MyLifeBrowser interface on web browser. The map image is provided by NaverMap API. The yellow marks mean the trajectory of mobile user and the blue marks show the position that the user has been listening mp3 music.

4) *Time period selection*: When the 'Instant Mode' button is checked, the 24 time period check boxes are appeared as shown in Fig. 7. They determine whether the log data of selected time period are seen or not.

5) *Media visualization*: The thumbnail images of photos are listed on the lower right area. When a thumbnail is clicked, the original size of image is shown on a new window.

## 5. Concept-based searching

With the MyLifeBrowser, we can also search the specific event based on concept. The method expands an input query concept by adding keywords related to it by using ConceptNet and Bayesian networks. The proposed method consists of the next three steps.

– *Concept expansion using Bayesian networks*: Adding keywords that have high linkage strength by analyzing the BNs.
– *Concept expansion using ConceptNet*: Adding keywords that have high relationship score by using the ConceptNet.
– *Searching*: Searching by using all of the expanded query keywords.

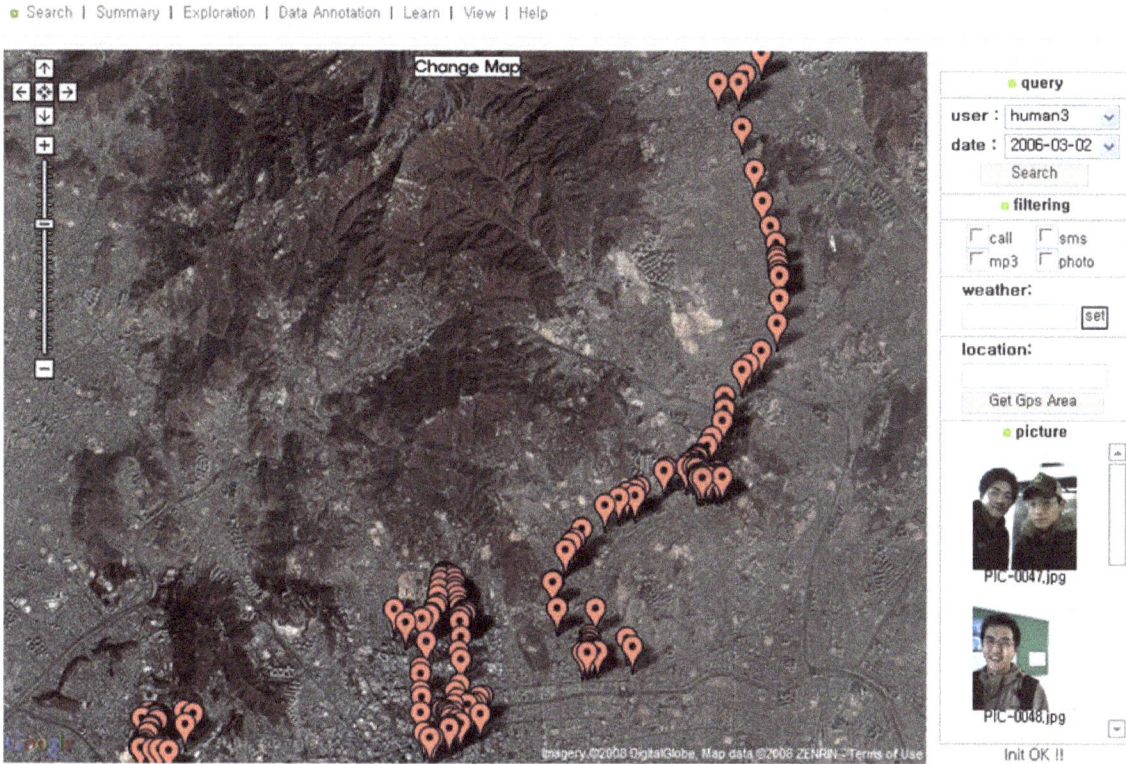

Fig. 6. MyLifeBrowser interface under satellite mode. The satellite map image is provided by Google Map API.

Fig. 7. Check boxes for time period selection.

### 5.1. Concept expansion using Bayesian networks

When the query keyword given exists in Bayesian networks, the other keywords related to the query keyword can be found by analyzing the structure of the Bayesian network. We exploit the Bayesian networks used in the high-level context reasoning module.

The strength between contexts and lifelogs defined in the Bayesian networks can be calculated by computing Noisy-OR strength value [29]. Noisy-OR strength is a parameter of CPT on Noisy-OR gate model, which is an approximated CPT computation method by canonical interaction models that require fewer parameters. The proposed Noisy-OR linkage strength $S_i$ is calculated as follows,

$$S_i = (p_i/0.5) - 1.0 \qquad (6)$$

where $p_i$ denotes a conditional probability when the cause $x_i$ and effect $y$ are activated (see the Eq. (7)).

$$p_i = \Pr(y|\bar{x}_1, \bar{x}_2, \ldots, x_i, \ldots, \bar{x}_{n-1}, \bar{x}_n) \qquad (7)$$

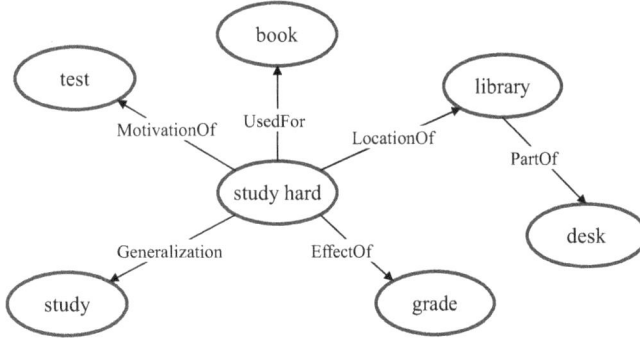

Fig. 8. An example of ConceptNet.

```
for (int a = 0; a<m; a++) {
  R_a = 0;
  R_a += w_a × Calculate_relationship(QA, L_a);
  R_a += w_c × Calculate_relationship(QC, L_a);
  R_a += w_d × Calculate_relationship(QD, L_a);
  R_a += w_s × Calculate_relationship(QS, L_a);
}

float Calculate_relationship(CQ, L_a) {
  float Value = 0.0f;
  for each(i ∈ CQ)
   for each(j ∈ L_a)
    if (cq_i == l^a_j)
     Value += cq_i;
  return Value;
}
```

Fig. 9. Concept expansion algorithm by ConceptNet toolkit.

The conversion from normal CPTs to Noisy-OR CPT is performed by the BN library, SMILE (Structural Modeling, Inference, and Learning Engine) 1.0 [30].

### *5.2. Concept expansion using ConceptNet*

ConceptNet developed by MIT Media Laboratory is a very-large semantic network of common sense knowledge and practical textual reasoning toolkit [31]. It aggregates 1,600,000 common sense knowledge extracted by heuristic rules from 700,000 sentences of Open Mind Common Sense (OMCS) project. The knowledge elements in ConceptNet are linked with temporal, spatial, physical, social, and mental relationships like Fig. 8. Since it contains sufficient general information, it is useful to develop intelligent system and conversational agent.

In order to find the concepts related to the query keyword from user log, by using the functions of ConceptNet Toolkit [28]. We searched the expanded keywords set $CQ = QA \cup QC \cup QD \cup QS$ for the query $Q = \{q_1, q_2, \ldots, q_n\}$. The $QA$, $QC$, $QD$, and $QS$ mean the project affective, project consequences, project details, and project spatial function of ConceptNet toolkit.

- *Project affective*: Extracting the concepts set $QA = \{a_1, a_2, \ldots, a_i\}$ emotionally similar to input concepts.
- *Project consequences*: Extracting the concepts set $QC = \{c_1, c_2, \ldots, c_j\}$ concurrently or successively related to input concepts.
- *Project details*: Extracting the concepts set $QD = \{d_1, d_2, \ldots, d_k\}$ similar features or lower event concepts to input concepts.
- *Project spatial*: Extracting the set of places $QS = \{s_1, s_2, \ldots, s_l\}$ related to the input concepts.

The correlation value $R$ between $Q$ and $L_t$ is calculated by comparing the concepts obtained by $CQ$ and lifelogs $\{L_1, L_2, \ldots, L_m\}$ ($m$ is the number of log data) like Fig. 9 and the expanded keywords selection is conducted based on the rank threshold for it.

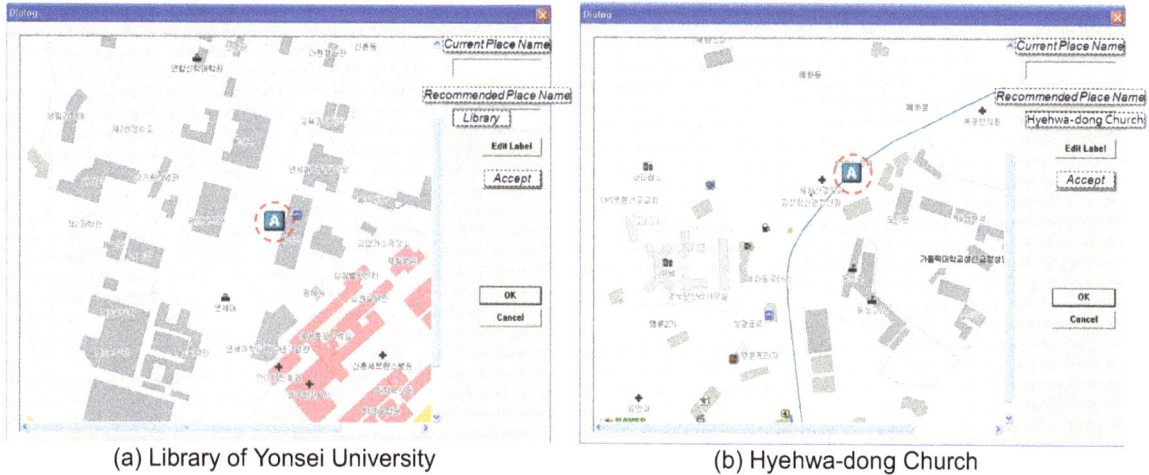

(a) Library of Yonsei University       (b) Hyehwa-dong Church

Fig. 10. The examples of location recommendation. The recommend place name came from open-source API. The current place name is the place name stored in DB. The accept button enables the DB update with the recommend place name.

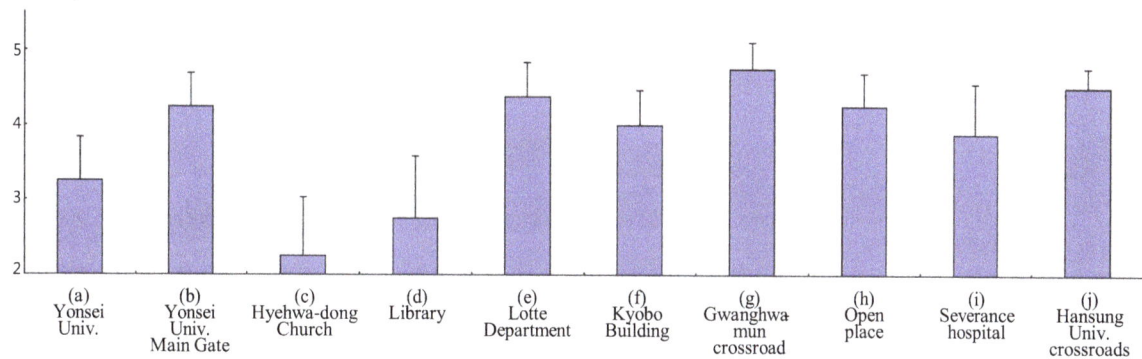

Fig. 11. The result of correctness survey for the distance based location name recommendation.

## 6. Experiments

In this section, we analyze the proposed browser in terms of experiments with collected experimental data. We gathered mobile lifelog data from three users and stored the data in the relational database and conducted experiments with them. We evaluate the location positioning system and the concept-based searching capability. In the experiments, Redhat 9.0 is used for the operating system of database server, and MySQL is used for database. The webpage was constructed by PHP and JavaScripts.

### 6.1. Performance evaluation of Location Positioning System

To test the Location Positioning System, we surveyed about the performance with eight graduate students. The GPS data for test were collected by college students for a day. We selected 10 points and evaluated the correctness of the proposed location labeling method. Figure 10 shows the result of the two places; 'Library of Yonsei University' and 'Hyehwa-dong Church'. The alphabet 'A' mark denotes the current position and the recommended place name is described on the text box at the right side. The

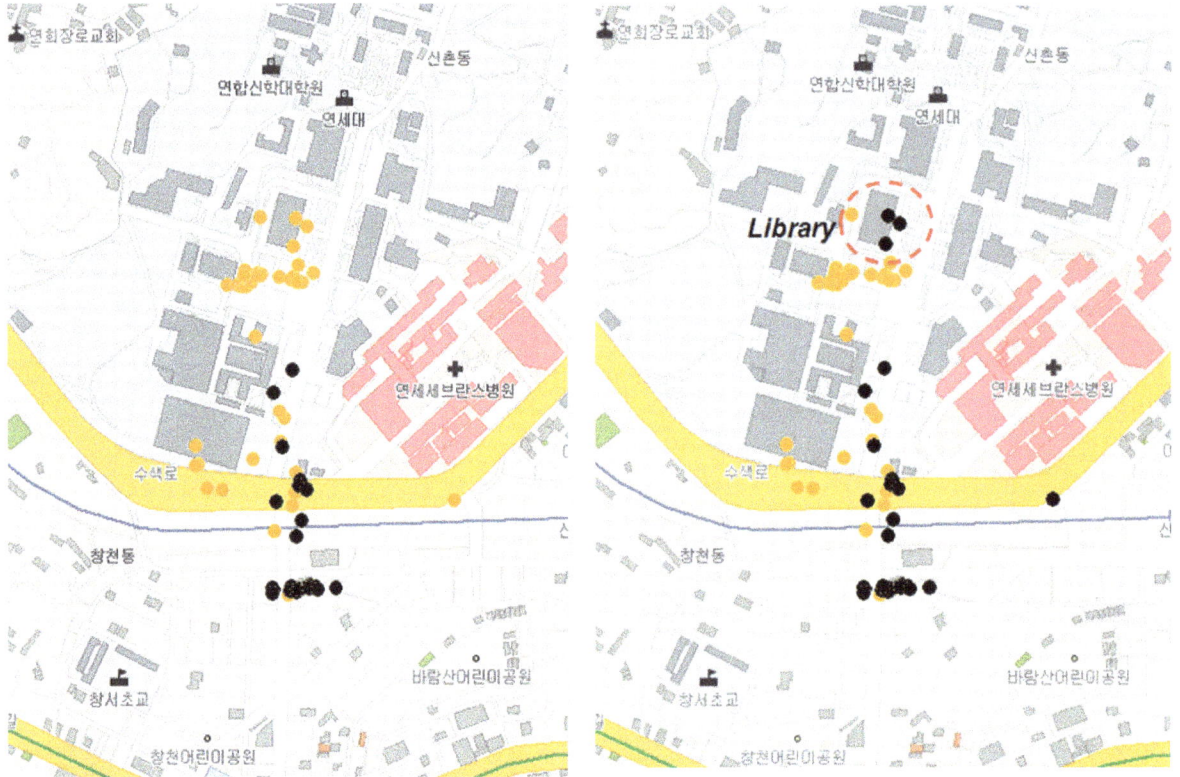

Fig. 12. The result of searching using a query keyword 'study'. The result of normal search and the concept-based search are shown at the left and right sides.

subjects were given the result of location recommendation for the 10 test places and provided the score of the correctness value from 1 to 5.

Figure 11 shows the mean and standard deviation of the evaluation values. The seven places excluding (a), (c), and (d) got the points equal to or larger than 4, since the areas of excepted places are wide and not in a circle. In more detail, we can discover that three places such as 'The library in Yonsei University', 'Yonsei University', and 'Hyehwa-dong Church' have relatively low scores for correctness value. Two places ('The library in Yonsei University' and 'Yonsei University') among them have spatial relationship with each other. Subjects in the experiment are undergraduate students of Yonsei University and know most areas in the university very well. Therefore, they are very sensitive about subtle labeling errors for these places. If locations near gates of the university can be determined as 'university', they will consider that the result is not appropriate for 'university'. On the other hand, 'Hyehwa-dong Church' cannot practically be mapped to church with GPS coordinates. It can be represented as 'road' close to church. In real life, they passed by 'Hyehwa-dong church' in a bus on their way to university. Distance-based automatic location labeling method often includes errors like them. However, the areas with low points can also be fixed by the proposed block-based manual labeling method.

## 6.2. Evaluation of concept-based search capability

To show an example of concept-based searching, we ran the search capability using the query keyword 'study' with log data collected on Feb. 25th by human1. The search result will help the user to find the

Table 2
The BN nodes related to the keyword 'study'

| BN name | Node name | Related node name | Link strength | Ranking |
|---|---|---|---|---|
| Music.BN | Study | Study with music | 0.80 | – |
| School.BN | Study | Study area | 0.68 | – |
| School.BN | Study | Study alone | 0.60 | – |
| Busy.BN | Study until late | Library | 0.52 | 1 |
| School.BN | Study area | Library | 0.50 | |
| Music.BN | Study | College building | 0.46 | 2 |
| Music.BN | Study | Library | 0.46 | |
| Busy.BN | Study until late | Busy | 0.36 | 3 |
| School.BN | Study area | High school | 0.00 | 4 |
| School.BN | Study area | School area | 0.00 | 4 |
| School.BN | Study area | Home | 0.00 | 4 |

Table 3
Top 6 keywords expanded from the query keyword 'study' by ConceptNet

| Query keyword | Expanded keyword | Relationship |
|---|---|---|
| Study | University | 0.04 |
| | Library | 0.04 |
| | School | 0.03 |
| | In house | 0.01 |
| | Room | 0.01 |
| | China | 0.00 |

Table 4
The keywords expanded by the proposed concept-based searching method

| Input | Expanded keyword |
|---|---|
| Meeting | Happy call, happy photo, free time, eat out, meet, watch, cooking, dishwashing, food photo, wash a face, meal (Korean), meal |
| Study | Lecture area, lecture, lecture time, study alone, study area, study, study with music, study until late, drowse place |
| Travel | Riding vehicle, good scenery place, scenery photo, bother, spare time, drowse, traffic jam |
| Meal | Place for meal, eat out, meal (Korean), meal, cooking, dishwashing |
| Reading | Study alone, study area, study, study with music, study until late, lecture time, lecture area, lecture |
| Lunch | Take a tea, place for meal, eat out, meal (Korean), meal |

lifelog and contexts related to the 'study' activity. Figure 12 shows the result. The left side shows the result of normal searching and the right side shows that of concept-based searching, and black dots show the lifelogs that are related to keyword 'study'. We can see that the black marks are added around the library in the result of the proposed method.

The expanded keywords by the process of concept-based searching are listed in the Tables 2 and 3. Table 2 shows the search result using the keywords expanded by Bayesian networks. The keywords were extracted from the high-level context nodes linked to the nodes related directly to the query keyword 'study. ' The ranking bound for selection was 4. Table 3 shows the search result using the keywords expanded by ConceptNet. We can see that the keywords 'university', 'library', and 'school' can be expanded with the ranking parameter 4.

To evaluate the usability of concept-based searching capability, we collected a month of log data from three college students, and analyzed the lifelog data with their activity diaries, and conducted search experiments with six keywords; meeting, study, travel, meal, reading, and lunch. These six keywords are proposed by the authors. 'Meal' and 'lunch' among them are similar keywords but have little different

Table 5
A part of results of concept-based search listed by date

| Human | Date | Meeting | Study | Travel | Meal | Reading | Lunch | # of better results |
|-------|------|---------|-------|--------|------|---------|-------|---------------------|
| h1 | 2/24 | X | − | X | X | X | X | 0 |
|    | 2/27 | X | + | X | X | X | X | 1 |
|    | 2/28 | X | + | X | X | X | X | 1 |
|    | 3/01 | + | X | X | X | X | X | 1 |
|    | 3/02 | + | + | X | + | + | + | 5 |
|    |      |   |   | ... |   |   |   |   |
| h2 | 2/24 | − | X | X | X | X | + | 1 |
|    | 2/25 | + | − | + | + | − | + | 4 |
|    | 2/28 | X | X | X | X | X | X | 0 |
|    | 3/05 | X | X | X | X | X | X | 0 |
|    | 3/06 | X | X | X | X | X | X | 0 |
|    |      |   |   | ... |   |   |   |   |
| h3 | 2/25 | + | + | + | + | + | + | 6 |
|    | 2/26 | X | + | X | X | X | X | 1 |
|    | 2/28 | X | − | X | X | X | X | 0 |
|    | 3/01 | + | X | + | X | X | X | 2 |
|    | 3/02 | + | + | + | X | + | + | 5 |
|    |      |   |   | ... |   |   |   |   |

Table 6
The results of concept-based search

|  | Meeting | Study | Travel | Meal | Reading | Lunch | Total |
|--|---------|-------|--------|------|---------|-------|-------|
| Test | 32 | 32 | 32 | 32 | 32 | 32 | 192 |
| Change | 21 | 24 | 14 | 13 | 15 | 17 | 104 |
| Better | 11 | 19 | 9 | 13 | 13 | 17 | 82 |
| Change rate | 0.66 | 0.75 | 0.44 | 0.41 | 0.47 | 0.53 | 0.54 |
| Better change rate | 0.52 | 0.79 | 0.64 | 1.00 | 0.87 | 1.00 | 0.79 |

meanings in this paper. 'Meal' means eating some food regardless of time but 'lunch' is eating food during lunch time. As 'meal' is a more general concept than 'lunch', the result for searching 'meal' will also include the result of searching 'lunch'. In this process, we used Korean-English conversion table to find the matched location name.

The keywords expanded by the proposed concept-based searching method are shown in Table 4. The experimental results for the six query keywords for whole date are shown in Table 5. It shows that the result for each day is different. In the table, 'X' means no change, '+' denotes better result that means the expanded marks are relative to the event in real, '−' denotes worse result, which is incorrect changes. We assumed that the 'travel' includes strange place and the 'meeting' includes all meeting between friends. The Table 5 shows that the improvement rate of searching capability is 42% as the number of searching is 192 and the number of better results is 19.

Table 6 shows whether the changed results of Table 5 is better or not. The number of changes was 104 and that of better results was 82, so the improvement rate by concept-based search was 82%. The results for 'meal', 'study', and 'reading' events are especially better because they were relatively frequent events and they had more related events or high-level contexts.

## 7. Concluding remarks

In this paper, we developed a mobile life browser by constructing a database from the lifelog data collected on mobile devices, analyzing and summarizing the log data and visualizing the personal everyday-life effectively. For visualization in maps, we exploited open-source APIs; NaverMap and Google Map. The mobile life browser visualized the lifelog data on the map-based interface and managed personal everyday-life easily. It also supported concept-based searching capability that was searching related concepts and log data based on common sense database by ConceptNet and contexts probability distribution by Bayesian networks. The experiments confirmed the usefulness of the mobile life browser and the concept-based searching capability.

However, since the system including the open-source API of Korea worked in Korean while the ConceptNet Toolkit operated in English, the conversion between English and Korean was time consuming and wasteful. In the near future, we will have to unify the language bases. Also, experiments with sufficient real world data should be conducted for a longer period of time. The subjective test shows the accuracy and effectiveness of the proposed browser. However, it cannot reveal the usefulness and usability of it completely and should be conducted in the future in order to deal with user's difficulties while using system. Finally, privacy issues are very important part of the system using personal information like this. In order to prevent invasion of privacy, personal agreement and authentication are required at all processes such as collecting life log, seeing life log and searching with keywords. Our future work will consider these privacy issues and policies.

## Acknowledgments

This work was supported by the industrial strategic technology development program, 10044828, development of augmenting multisensory technology for enhancing significant effect on service industry, funded by the Ministry of Trade, Industry and Energy (MI, Korea).

## References

[1] Nokia LifeBlog, http://www.nokia.com/lifeblog.
[2] A.K. Dey, Understanding and using context, *Personal Ubiquitous Computing* **5**(1) (2001), 4–7.
[3] B. Schilit, N. Adams and R. Want, Context-aware computing applications, *IEEE Workshop on Mobile Computing Systems and Applications*, Santa Cruz, CA, US, (1994), 89–101.
[4] A. Schmidt, A. Takaluoma and J. Mntyjrvi, Context-aware telephony over WAP, *Personal Technologies* **4**(4) (2000), 225–229.
[5] B.P.L. Lo, S. Thiemjarus and G.-Z. Yang, Adaptive Bayesian networks for video processing, *Int Conf on Image Processing* **1**(1) (2003), 889–892.
[6] S. Sweeney and F. Crestani, Effective search results summary size and device screen size: Is there a relationship? *Information Processing and Management* **42** (2006), 1056–1074.
[7] J. Otterbacher, D. Radev and O. Kareem, Hierarchical summarization for delivering information to mobile devices, *Information Processing and Management* **44** (2008), 931–947.
[8] M. Raento, A. Oulasvirta, R. Petit and H. Toivonen, ContextPhone: A prototyping platform for context-aware mobile applications, *IEEE Pervasive Computing* **4**(2) (2005), 51–59.
[9] A. Krause, A. Smailagic and D.P. Siewiorek, Context-aware mobile computing: Learning context-dependent personal preferences from a wearable sensor array, *IEEE Trans on Mobile Computing* **5**(2) (2006), 113–127.
[10] R. DeVaul, M. Sung, J. Gips and A. Pentland, MIThril 2003: Applications and architecture, *Proc of 7th IEEE Int Symposium on Wearable Computers* (2003), 4–11.
[11] P. Zheng and L.M. Ni, The rise of the smart phone, *IEEE Distributed Systems Online* **7**(3) (2006).
[12] P. Korpipaa, J. Mantyjarvi, J. Kela, H. Keranen and E.-J. Malm, Managing context information in mobile devices, *IEEE Pervasive Computing* **2**(3) (2003), 42–51.
[13] J. Gemmell, L. Williams, K. Wood, R. Lueder and G. Bell, Pervasive capture and ensuing issues for a personal lifetime store, *Proc of the 1st ACM Workshop on Continuous Archival and Retrieval of Personal Experiences* (Oct 2004), 48–55.

[14]   D.P. Siewiorek, A. Smailagic, J. Furakawa, A. Krause, N. Moraveji, K. Reiger, J. Shaffer and F.L. Wong, SenSay: A context-aware mobile phone, *Proc 7th Int Symp of Wearable Computers* (Oct 2003), 248–249.

[15]   A.F. Smeaton, Content vs, context for multimedia semantics: The case of SenseCam image structuring, *Lecture Notes in Computer Science* **4306** (2006), 1–10.

[16]   A. Schmidt, Ubiquitous computing – computing in context, Ph.D. Dissertation, Lancaster University, 2003.

[17]   M. Raento, A. Oulasvirta, R. Petit and H. Toivonen, ContextPhone: A prototyping platform for context-aware mobile applications, *IEEE Pervasive Computing* (2005), 51–59.

[18]   S.N. Eagle and A. Pentland, Social serendipity: Mobilizing social software, *IEEE Pervasive Computing* (April–June 2005), 28–34.

[19]   J. Gemmell, G. Bell and R. Lueder, MyLifeBits: A personal database for everything, *Communications of the ACM* **49**(1) (January 2006).

[20]   E. Horvitz, S. Dumais and P. Koch, Learning predictive models of memory landmarks, *26th Annual Meeting of the Cognitive Science Society* (2004), 1–6.

[21]   N. Eagle, Machine perception and learning of complex social systems, Ph.D. Thesis, Massachusetts Institute of Technology, 2005.

[22]   S.-B. Cho, K.-J. Kim, K.-S. Hwang and I.-J. Song, AniDiary: Daily cartoon-style diary exploits Bayesian networks, *IEEE Pervasive Computing* (Jul–Sep 2007), 66–75.

[23]   Microsoft MapPoint, http://www.microsoft.com/mappoint/default.mspx.

[24]   J. Hightower, From position to place, *Proc of the 2003 Workshop on Location-Aware Computing* (2003), 10–12.

[25]   J. Hightower, S. Consolvo, A. LaMarca, I. Smith and J. Hughes, Learning and recognizing the places we go, *In Proceedings of the Seventh International Conference on Ubiquitous Computing* (Ubicomp 2005) Lecture Notes in Computer Science, (2005), 159–176.

[26]   C. Zhou, J.P. Ludford, D. Frankowski and G.L. Terveen, An experiment in discovering personally meaningful places from location data, in: *CHI Extended Abstracts* (2005), 2029–2032.

[27]   P. Nurmi and S. Bhattacharya, Identifying meaningful places: The non-parametric way, in: *Pervasive* (2008), 111–127.

[28]   K.B. Korb and A.E. Nicholson, *Bayesian Artificial Intelligence* Chapman & Hall/CRC, (2003).

[29]   K.-S. Hwang and S.-B. Cho, Modular Bayesian networks for inferring landmarks on mobile daily life, *The 19th Australian Joint Conf on Artificial Intelligence* (2006), 929–933.

[30]   Smile Library, http://genie.sis.pitt.edu.

[31]   H. Liu and P. Singh, ConceptNet – A practical commonsense reasoning tool-kit, *BT Technology Journal* **22**(4) (2004), 211–226.

**K.-S. Hwang** is a researcher at the Softcomputing Laboratory and a doctoral student in computer science at Yonsei University. His research interests include Bayesian networks and evolutionary algorithms for context-aware computing and intelligent agents. He received his MSc in computer science from Yonsei University.

**S.-B. Cho** received the B.S. degree in computer science from Yonsei University, Seoul, Korea, in 1988 and the M.S. and Ph.D. degrees in computer science from KAIST (Korea Advanced Institute of Science and Technology), Taejeon, Korea, in 1990 and 1993, respectively. He worked as a Member of the Research Staff at the Center for Artificial Intelligence Research at KAIST from 1991 to 1993. He was an Invited Researcher of Human Information Processing Research Laboratories at ATR (Advanced Telecommunications Research) Institute, Kyoto, Japan from 1993 to 1995, and a Visiting Scholar at University of New South Wales, Canberra, Australia in 1998. He was also a Visiting Professor at University of British Columbia, Vancouver, Canada from 2005 to 2006. Since 1995, he has been a Professor in the Department of Computer Science, Yonsei University. His research interests include neural networks, pattern recognition, intelligent man-machine interfaces, evolutionary computation, and artificial life. Dr. Cho was awarded outstanding paper prizes from the IEEE Korea Section in 1989 and 1992, and another one from the Korea Information Science Society in 1990. He was also the recipient of the Richard E. Merwin prize from the IEEE Computer Society in 1993. He was listed in Who's Who in Pattern Recognition from the International Association for Pattern Recognition in 1994, and received the best paper awards at International Conference on Soft Computing in 1996 and 1998. Also, he received the best paper award at World Automation Congress in 1998, and listed in Marquis Who's Who in Science and Engineering in 2000 and in Marquis Who's Who in the World in 2001. He is a Senior Member of IEEE and a Member of the Korea Information Science Society, INNS, the IEEE Computer Society, and the IEEE Systems, Man, and Cybernetics Society.

# 3

# MANET performance for source and destination moving scenarios considering OLSR and AODV protocols

Elis Kulla[a,*], Masahiro Hiyama[a], Makoto Ikeda[b], Leonard Barolli[c], Vladi Kolici[d] and Rozeta Miho[d]

[a]*Graduate School of Engineering, Fukuoka Institute of Technology (FIT), 3-30-1 Wajiro-Higashi, Higashi-Ku, Fukuoka, Japan*

[b]*Center for Asian and Pacific Studies, Seikei University, 3-3-1 Kichijoji-Kitamachi, Musashino-Shi, Tokyo, Japan*

[c]*Department of Information and Communication Engineering, Fukuoka Institute of Technology (FIT), 3-30-1 Wajiro-Higashi, Higashi-Ku, Fukuoka, Japan*

[d]*Department of Electronic and Telecommunication, Polytechnic University of Tirana, Mother Teresa Square, Nr. 4, Tirana, Albania*

**Abstract.** Recently, a great interest is shown in MANETs potential usage and applications in several fields such as military activities, rescue operations and time-critical applications. In this work, we implement and analyse a MANET testbed considering AODV and OLSR protocols for wireless multi-hop networking. We investigate the effect of mobility and topology changing in MANET and evaluate the performance of the network through experiments in a real environment. The performance assessment of our testbed is done considering throughput, number of dropped packets and delay. We designed four scenarios: Static, Source Moving, Destination Moving and Source-Destination Moving. From our experimental results, we concluded that when the communicating nodes are moving and the routes change quickly, OLSR (as a proactive protocol) performs better than AODV, which is a reactive protocol.

## 1. Introduction

During recent years, we have witnessed a lot of research on wireless networks [5,16,17,21,1,8,24,25]. There are two network architectures for wireless networks: infrastructure and ad-hoc architecture.

Wireless networks often extend, rather than replace, wired networks, which are referred to as infrastructure networks. The wide area and local area wired networks are used as the backbone network. The wired backbone connects to special switching nodes called Base Stations (BSs). The BSs are often conventional PCs and workstations equipped with custom wireless adapter cards. They are responsible for coordinating access to one or more transmission channel(s) for mobiles located within the coverage cell.

Ad-hoc networks, on the other hand, are multi-hop wireless networks in which a set of mobile nodes cooperatively maintain network connectivity. Ad-hoc networks are characterized by dynamic,

---

*Corresponding author. E-mail: eliskulla@yahoo.com.

unpredictable, random, multi-hop topologies with typically no infrastructure support. The mobile nodes must periodically exchange topology information which is used for routing updates.

A Mobile Ad hoc Network (MANET) is a collection of wireless mobile terminals that are able to dynamically form a temporary network without any aid from fixed infrastructure or centralized administration. In recent years, MANET are continuing to attract the attention for their potential use in several fields. Mobility and the absence of any fixed infrastructure make MANET very attractive for mobility and rescue operations and time-critical applications.

Most of the work for MANETs has been done in simulation, as in general, a simulator can give a quick and inexpensive understanding of protocols and algorithms [9,10,20,26]. However, experimentation in the real world are very important to verify the simulation results and to revise the models implemented in the simulator. A typical example of this approach has revealed many aspects of IEEE 802.11, like the gray-zones effect [14], which usually are not taken into account in standard simulators, as the well-known *ns-2* simulator [22].

So far, we can count a lot of computer simulation results on the performance of MANET, e.g. in terms of end-to-end throughput, delay and packet loss. However, in order to assess the computer simulation results, real-world experiments are needed and a lot of testbeds have been built to date [12,23,13]. The baseline criteria usually used in real-world experiments is guaranteeing the repeatability of tests, i.e. if the system does not change along the experiments. How to define a change in the system is not a trivial problem in MANET, especially if the nodes are mobile.

In this paper, we focus on comparing the performance of two types of routing protocols Ad-hoc On demand Distance Vector (AODV), which is a reactive routing protocol, and Optimized Link State Routing (OLSR), which is a proactive routing protocol. Both protocols have been gaining great attention within the scientific community. Furthermore, the *aodv-uu* [4] and the *olsrd* [18] software we have used in our experiments are the most updated software we have encountered.

In our previous work, we found the following results. We proved that while some of the OLSR's problems can be solved, for instance the routing loop, this protocol still have the self-interference problem. There is an intricate inter-dependence between MAC layer and routing layer, which can lead the experimenter to misunderstand the results of the experiments. We carried out the experiments considering stationary nodes of ad-hoc network. We considered the node mobility and carry out experiments for AODV, OLSR and BATMAN protocols [2]. We found that throughput of TCP were improved by reducing Link Quality Window Size (LQWS), but there were packet loss because of experimental environment and traffic interference. For TCP data flow, we got better results when the LQWS value was 10.

In this work, we implemented four MANET scenarios and carried out real world experiments in an indoor environment. We assess the performance of two routing protocols AODV and OLSR for different source and destination moving scenarios.

The structure of the paper is as follows. In Section 2, we give a short description of AODV and OLSR. In Section 3, we describe the testbed and its implementation. The moving scenarios are described in Section 4. In Section 5, we present experimental evaluation. Finally, conclusions are given in Section 6.

## 2. Routing protocols

### 2.1. AODV overview

AODV is one of the most popular reactive routing protocol for MANETs [19]. As a reactive (on demand) protocol, when a node wants to transmit data, it first starts a route discovery process, by flooding a RREQ (Route Request) packet. The RREQ packets are forwarded by all nodes by which they are received. This procedure continues until the destination is found. On the way to destination, the RREQ informs all the intermediate nodes about a route to the source. When the RREQ reaches the destination, destination sends a Route Reply (RREP) packet which follows the reverse path discovered by RREQ. This informs all intermediate nodes about a route to the destination node. After RREQ and RREP are delivered to their destination, each intermediate node on the route knows what node to forward data packets in order to reach source or destination. Thus data packets do not need to carry addresses of all intermediate nodes in the route. It just carries the address of the destination node, decreasing noticeably routing overheads.

A third kind of routing message, called Route Error (RERR), allows nodes to notify errors, for example, because a previous neighbor has moved and is no longer reachable. If the route is not active (i.e., there is no data traffic flowing through it), all routing information expire after a timeout and is removed from the routing table.

In AODV, the route discovery process may last for a long time, or it can be repeated several times, due to potential failures during the process. This introduces extra delays, and consumes more bandwidth as the size of the network increases.

## 2.2. OLSR overview

The link state routing protocol that is most popular today in the open source world is OLSR from olsr.org. OLSR with Link Quality (LQ) extension and fisheye-algorithm works quite well. The OLSR protocol is a pro-active routing protocol, which builds up a route for data transmission by maintaining a routing table inside every node of the network.

The routing table is computed upon the knowledge of topology information, which is exchanged by means of Topology Control (TC) packets. The TC packets in turn are built after every node has filled its neighbors list. This list contains the identity of neighbor nodes. A node is considered a neighbor if and only if it can be reached via a bi-directional link. OLSR checks the symmetry of neighbors by means of a 4-way handshake based on the so called HELLO messages. This handshake is inherently used to compute the packet loss probability over a certain link. This can sound odd, because packet loss is generally computed at higher layer than routing one. However, an estimate of the packet loss is needed by OLSR in order to assign a weight or a state to every link.

In OLSR, control packets are flooded within the network by electing special nodes, called Multi Point Relays (MPRs), to the role of forwarding nodes. By this way, the amount of control traffic can be reduced. These nodes are chosen in such a way that every node can reach its neighbors 2-hops far away. In our OLSR code, a simple RFC-compliant heuristic is used [3] to compute the MPR nodes. Every node computes the path towards a destination by means of a simple shortest-path algorithm, with hop-count as target metric. In this way, a shortest path can result to be also not good, from the point of view of the packet error rate. Accordingly, recently OLSRd has been equipped with the Link Quality (LQ) extension, which is a shortest-path algorithm with the average of the packet error rate as metric. This metric is commonly called as the Expected Transmission Rate (ETX), which is defined as $\text{ETX}(i) = 1/(NI(i) \times LQI(i))$. Given a sampling window $W$, $\text{NI}(i)$ is the packet loss probability seen by a node on the $i$-th link during $W$. Similarly, $\text{LQI}(i)$ is the estimation of the packet loss seen by the neighbor node which uses the $i$-th link. When the link has a low packet error rate, the ETX metric is higher. The LQ extension greatly enhances the packet delivery ratio with respect to the hysteresis-based technique [6].

Table 1
Node addressing table

| Node ID | IP Address | Operating System |
|---------|------------|------------------|
| Node 1 | 192.168.0.1 | Fedora Core 4 |
| Node 2 | 192.168.0.2 | Ubuntu 9.04 |
| Node 3 | 192.168.0.5 | Ubuntu 9.04 |
| Node 4 | 192.168.0.6 | eeeUbuntu 9.04 |
| Node 5 | 192.168.0.7 | Ubuntu 9.04 |
| Node 6 | 192.168.0.10 | Ubuntu 9.04 |
| Node 7 | 192.168.0.11 | Ubuntu 9.04 |

(a) Node 1                                           (b) Node 2

Fig. 1. Hardware of the testbed.

## 3. Testbed description

### 3.1. Testbed environment

We implemented a MANET testbed and carried out experiments in the fifth floor of Building D, at Fukuoka Institute of Technology. This testbed provides the environment to make different measurements for indoor and outdoor communications. However, in this paper we deal only with indoor environment.

### 3.2. Operating system and routing software

The operating system installed on machines is Ubuntu 9.04 Linux (x5), eeeUbuntu 9.04 Linux (x1) all with kernel 2.6.28-18-generic and Fedora Core 4 Linux (x1) as shown in Table 1. Each of them can support all installed routing softwares.

In each machine, the AODV and OLSR routing softwares are installed from their source code in their respective web pages. Both of them are open source. See [4,18] for more information.

### 3.2.1. Network configuration

All machines used their own wireless adapter, except for the Fedora machine which uses a Linksys wireless card, whose drivers can be found at [15]. Each machine wireless card transmits at frequency 2.412 GHz (channel 1), and is put to ad-hoc infrastructure. In Fig. 1, we show a screen-shot of every node we used in experiments. Node IDs and IP addresses are shown in Table 1.

Fig. 2. GUI interface for parameter settings.

### 3.2.2. Traffic generation and getting the data

After configuring the network all nodes are put to their respective position, in accordance to the experimental scenario. To generate some traffic between nodes, we used D-ITG (Distributed Internet Traffic Generator) software, which is a Traffic Generator [7]. With D-ITG, one could send different type of traffics from one node to another. The amount of information to be sent and the duration of the transmission is set as an option. After finishing the transmission, D-ITG offers decoding tools to get information about network metrics along the whole transmission duration.

### 3.2.3. Testbed interface

All settings, editing and calculations can be done with the aid of a Graphical User Interface (GUI) as shown in [11]. This is helpful in saving time in the case of repeated experiments, and avoiding misprints during set-up. The GUI uses wxWidgets tool and each operation is implemented by Perl language. wxWidgets is a cross-platform GUI and tools library for GTK, MS Windows and Mac OS X. Many parameters are implemented in the interface such as transmission duration, number of trials, source address, destination address, packet rate, packet size, LQWS, and topology setting function. These parameters can be saved in a text file and can manage the experimental conditions in a better approach. The GUI interface of the implemented testbed is shown in Fig. 2.

## 4. Topology description

The implemented testbed provides a real-time system for analysing various aspects of MANETs. The purpose of this paper is to evaluate the performance of two routing protocols: AODV and OLSR. Performance evaluation is done for four different scenarios. The MAC filtering is not used in these experiments, so the nodes form e Mesh Topology. We describe the four scenarios in the following. The

Table 2
Experimental parameters

| Parameters | SS | SMS | DMS | SDMS |
|---|---|---|---|---|
| Nr. of experiments | 20 | 10 | 10 | 10 |
| Duration of experiment(s) | 60 | 120 | 120 | 120 |
| Packet rate (pkt/s) | 200 | 200 | 200 | 200 |
| Packet size (bytes) | 512 | 512 | 512 | 512 |

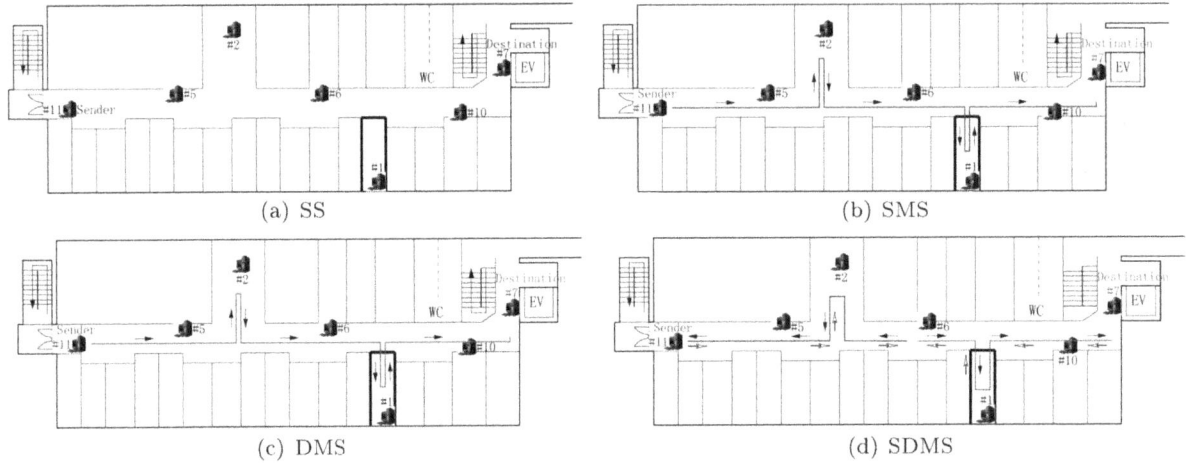

(a) SS

(b) SMS

(c) DMS

(d) SDMS

Fig. 3. Different topologies for experiments.

topologies for different experiments are shown in Fig. 3. All experimental parameters are shown in Table 2.

For the static scenario, 20 experiments were performed for each protocol, and every experiment lasted 60 seconds. The source node sent 512-byte packets, with a frequency of 200 packets per second. For the moving scenarios, we performed 10 experiments with a duration of 120 seconds each.

### 4.1. Static scenario

In the Static Scenario (SS), first all nodes are put in the positions shown in Fig. 3(a). Then, in each machine, the routing protocol deamons are started. In this paper, we consider AODV and OLSR and their deamons *aodvd* and *olsrd*, respectively. To let the routing protocol initialize routes, no data was transmitted for the first five minutes.

### 4.2. Source moving scenario

The Source Moving Scenario (SMS) is shown in Fig. 3(b). The nodes are in the same position as in SS (Fig. 3(a)), except that source node moved towards the destination node, as shown in Fig. 3(b). This movement is realized using a simple wheeled office chair.

### 4.3. Destination moving scenario

In Fig. 3(c), we show the Destination Moving Scenario (DMS). The destination node moves away from the source, starting its movement in the same position as the source node. At the end of 120 seconds, destination node and source node have the maximum distance between them.

Table 3
Average values for different experiments

| Nr. | Scenario | Protocol | Bitrate | Delay | Packetloss |
|-----|----------|----------|---------|-------|------------|
| 1 | SS | AODV | 819.1863 | 0.0032 | 0.000076 |
| 2 | | OLSR | 819.1727 | 0.0036 | 0.000056 |
| 3 | SMS | AODV | 613.9733 | 1.5855 | 0.2942 |
| 4 | | OLSR | 618.8715 | 1.6504 | 0.2532 |
| 5 | DMS | AODV | 720.2372 | 0.7445 | 0.1654 |
| 6 | | OLSR | 719.2644 | 0.8486 | 0.1597 |
| 7 | SDMS | AODV | 727.7739 | 0.8986 | 0.2265 |
| 8 | | OLSR | 775.7824 | 0.8352 | 0.1656 |

### 4.4. Source-destination moving scenario

As shown in Fig. 3(d), in Source-Destination Moving Scenario (SDMS), both source node and destination nodes are moving. Starting near the position of node 6, they both move away from each other for the first 60 seconds. Then they go back by the same route, to the starting position for the last 60 seconds.

## 5. Performance evaluation

### 5.1. Experimental settings

We performed the experiments in indoor environment (our departmental floor), as shown in Fig. 3, with the size nearly 70 m $\times$ 25 m. We used UDP traffic for experimental environment (see Table 2). The D-ITG is used to create the traffic and to collect the data. Data in the network were collected in a Mesh Topology for different scenarios of node movement and for two routing protocols. We were interested in Bitrate (kbps), Delay (ms) and Packetloss (No.of packets).

We used CBR (Constant Bit Rate) over UDP to create the traffic. The transmission rate of the data flow is 200 pkts/s, and the packet size is fixed to 512 kB, meaning a maximum bitrate of 819.2 kbps. Nodes (laptops) could access each other within the 70 meter region where the experiments were performed. We checked this by the $ping$ command of Ubuntu 9.04. In total, we performed 8 experiments, as shown in Table 3.

As MAC protocol we used the IEEE 802.11 b protocol and configured the wireless cards to operate at central frequency 2.412 GHz (channel 1) and with enough power to have connectivity with every node in the network. The main interest on these experiments was in the routing protocols and their behaviour in different scenarios, so all MAC parameters were kept unchanged. We should mention that during experiments all the IEEE 802.11 spectrum had been used by other access points operating within the campus, causing a considerable interference.

We took samples of 500 ms for every experiment, and computed the averages of each sample, using linux bash scripting and Matlab.

### 5.2. Experimental measurements

In Table 3, we show all the calculated average values for every experiment. We investigated all mean values of Bitrate, Delay and Packetloss, which are measured in "kilobits per second (kbps)", "milliseconds (ms)" and "percentage (%)", respectively.

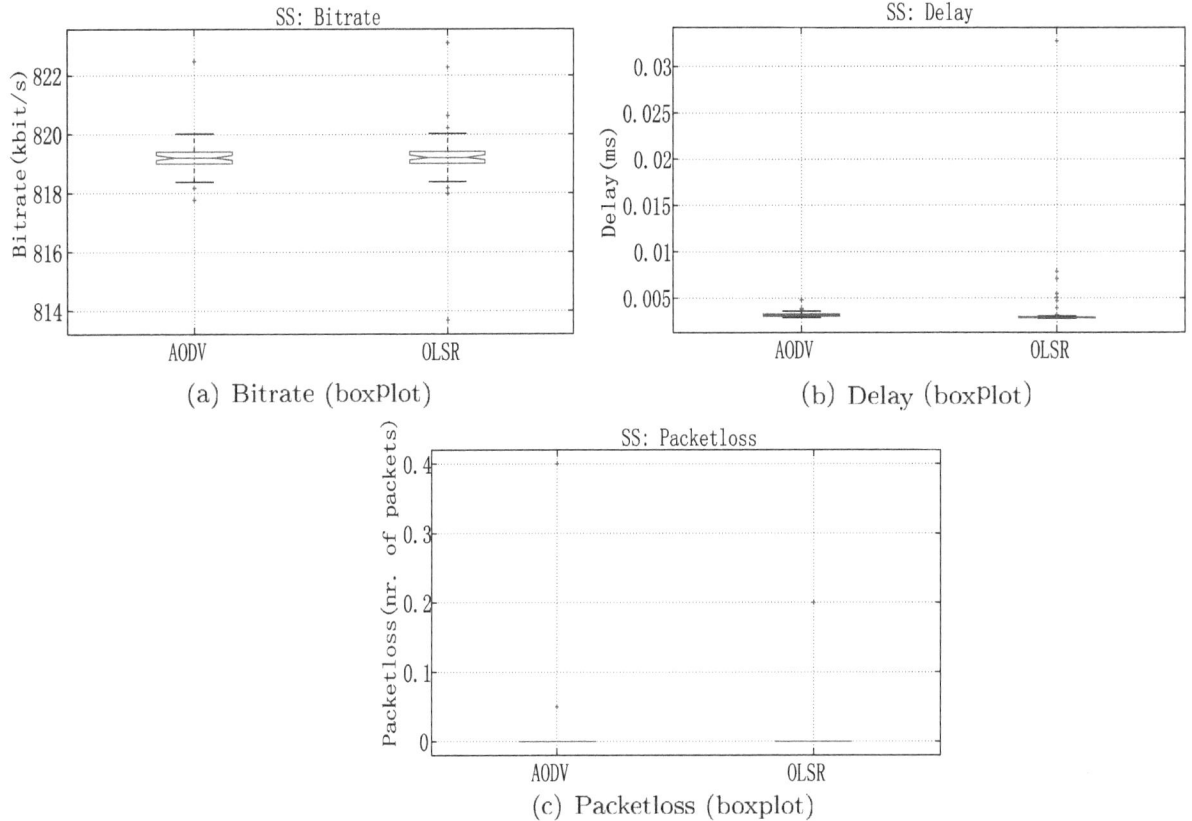

(a) Bitrate (boxPlot)

(b) Delay (boxPlot)

(c) Packetloss (boxplot)

Fig. 4. Different metrics vs different protocols for SS (boxplot).

For SS, in Fig. 4, we can see that for both AODV and OLSR, bitrate is almost the maximum (max = 819.2). This means that the routes have been established and the communication is performed at almost maximum performance. This is also shown in Table 3.

In SMS, the source node is approaching the destination node and at two time periods 30 s–50 s and 70 s–90 s they loose LOS (Line of Sight). In Figs 5 and 6, we show three metrics in time-domain and boxplot, respectively. In Fig. 5(a), the bitrate in the period of time 30 s–50 s reaches the value 0. This means that the source node could not find a route to the destination node. At this period of time the nodes loose LOS and a complete route (2 or more hops) is difficult to be established. At the time period 70 s–90 s, we also observe a decrease on the value of bitrate, which is more considerable in the case of AODV. In this case, even though there is no LOS between the communicating nodes, they are closer to each other and 1-hop or 2-hops routes can be quickly re-established.

In Fig. 6(a), we can observe that both protocols show the same performance regarding bitrate metric. At the period of time 30 s–50 s, when the bitrate reaches very low values, we notice a proportional increase in packetloss as shown in Fig. 5(c). At time period 70 s–90 s, we encountered a considerable amount of packet loss for AODV.

In Fig.6(c), it is shown that both protocols show almost the same performance considering packetloss metric. At time periods 30 s–50 s, in the case of OLSR, we notice that the delay is increased as shown in Fig. 5(b). At this time period, the communicating nodes are in NLOS (Non Line of Sight) conditions and the communication needs 2 or more hops to occur. Thus, as described in [2], OLSR performance

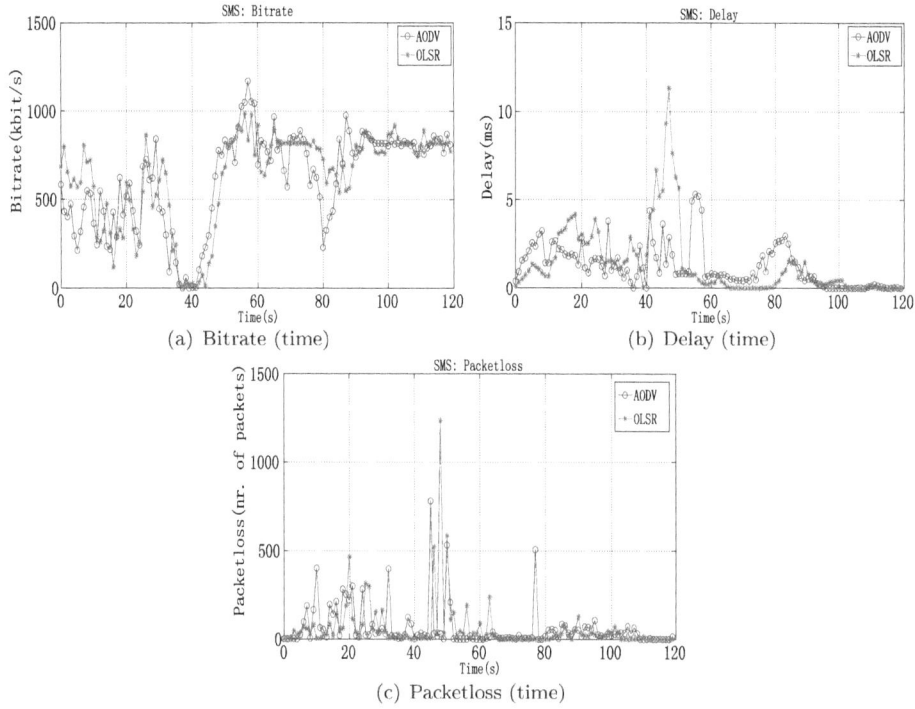

(a) Bitrate (time)

(b) Delay (time)

(c) Packetloss (time)

Fig. 5. Different metrics vs different protocols for SMS (time).

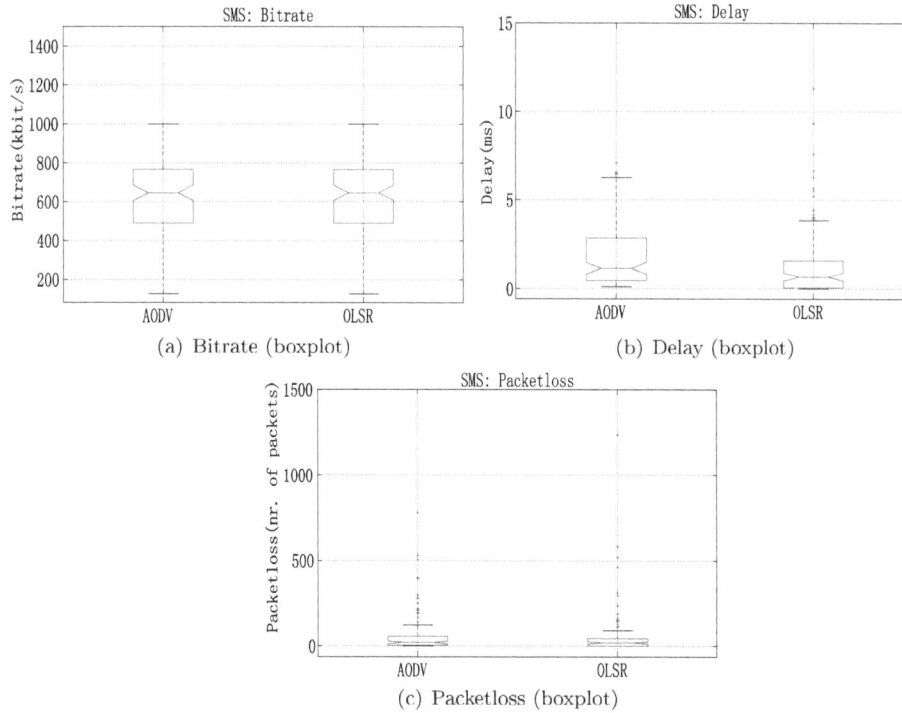

(a) Bitrate (boxplot)

(b) Delay (boxplot)

(c) Packetloss (boxplot)

Fig. 6. Different metrics vs different protocols for SMS (boxplot).

(a) Bitrate (time)

(b) Delay (time)

(c) Packetloss (time)

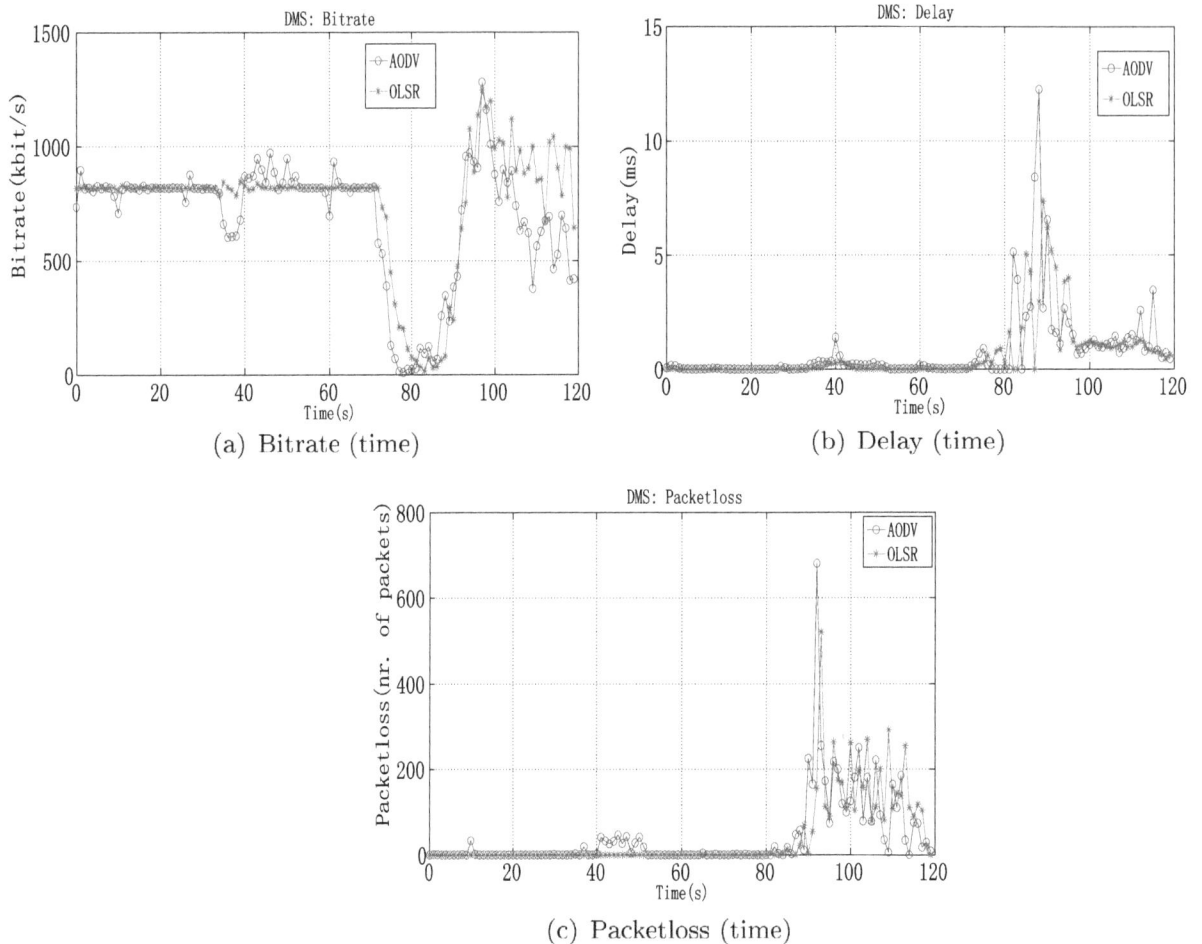

Fig. 7. Different metrics vs different protocols for DMS (time).

at 2-hops or 3-hops communication undergoes a degradation. As shown in Fig. 6(b), both OLSR and AODV protocols show the same performance considering the delay parameter.

In DMS, the destination node is moving away from the source node. In Figs 7 and 8, we show three metrics in time-domain and boxplot, respectively. In Fig. 7(a), the bitrate in the time period 75 s–90 s reaches the value 0, which means the source node could not find a route to the destination node. At this period of time the two nodes loose LOS and a complete route of 2 or more hops is difficult to be established. As is shown in Fig. 8(a), the OLSR has a better throughput than AODV. After time 90 s the bitrate in case of AODV is lower than the case when OLSR is used. This happens because at that time, routes need to be re-established, and for AODV the route discovery process is not always successful, thus it needs more time. This fact is reflected in delay and packetloss graphs, respectively in Figs 7(b) and 7(c) after time 90 s. At the period of time 75 s–90 s, when the bitrate reaches very low values, we notice a proportional increase in packetloss as shown in Fig. 7(c). After time 90 s the communications still has a considered amount of packetloss. In Fig. 8(c) is shown that AODV has a slightly worse performance than OLSR. At time period 75 s–90 s, we notice an increased delay in Fig.7(b), which is due to the low bitrate experienced at these time periods. As shown in Fig. 8(b), both AODV and OLSR protocols have almost the same performance.

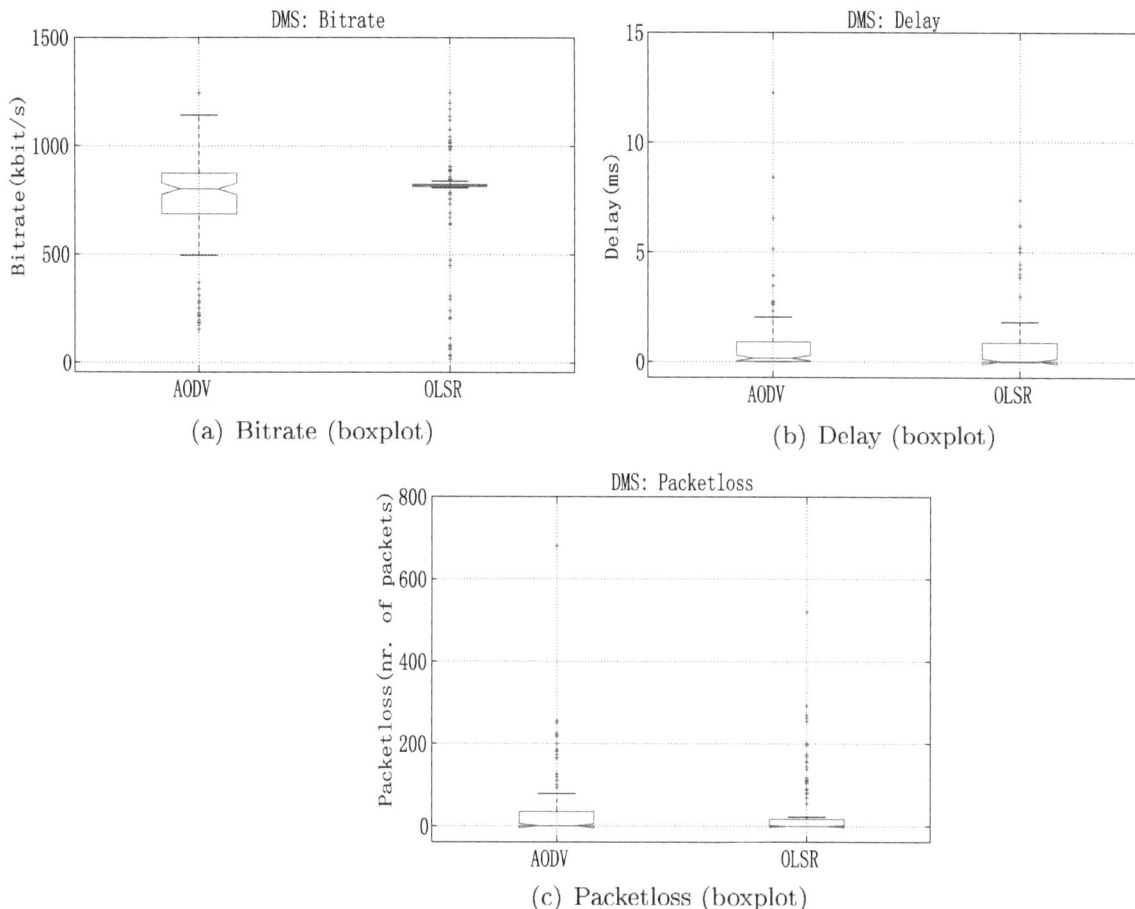

(a) Bitrate (boxplot)

(b) Delay (boxplot)

(c) Packetloss (boxplot)

Fig. 8. Different metrics vs different protocols for DMS (boxplot).

In SDMS during the first 60 seconds both nodes are moving away from each other and then during the last 60 seconds they are approaching each other via the same route of movement. In Figs 9 and 10, we show three metrics in time-domain and boxplot, respectively. In Fig. 9(a), the bitrate in the time periods 15 s–35 s and 90 s–105 s reaches the value 0, which means the source node could not find a route to the destination node. At this periods of time the nodes loose LOS and a complete route of 2 or more hops is difficult to be established. As it is shown in Fig. 10(a), OLSR has a better performance than AODV regarding bitrate metric. At time periods 15 s–35 s and 90 s–105 s when bitrate reaches very low values, we notice a proportional increase in packetloss as shown in Fig. 9(c). In Fig. 10(c) for packetloss metric, AODV has a slightly worse performance than OLSR. At time periods 15 s–35 s and 90 s–105 s, we notice an increased delay in Fig. 9(b), which is due to the low bitrate experienced at these time periods. As shown in Fig. 10(b), OLSR shows a better performance than AODV considering delay. This delay is caused by the continuous change of routes in SDMS.

AODV is more sensible to route changing, because it has to redefine the whole route before starting to send data. AODV protocol acts worse than OLSR in the cases when routes are lost. Being a reactive protocol, AODV has to redefine the communicating route, so it takes more time to re-establish the communication. In contrary OLSR chooses one of the old available routes, until the new routes are defined.

(a) Bitrate (time)

(b) Delay (time)

(c) Packetloss (time)

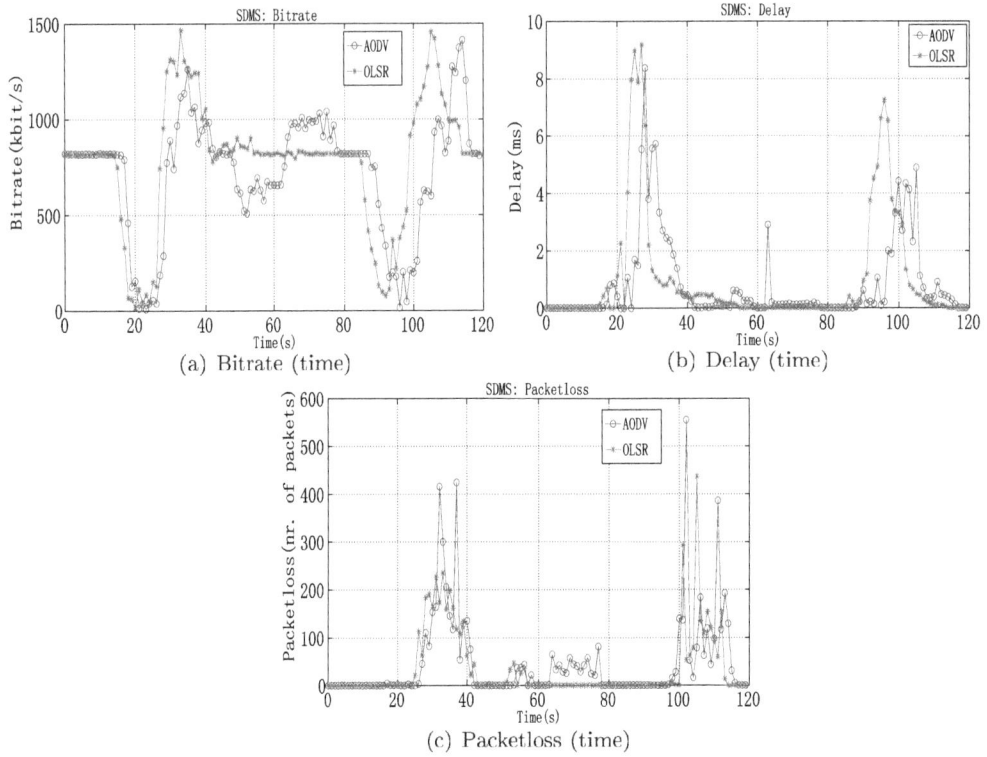

Fig. 9. Different metrics vs different protocols for SDMS (time).

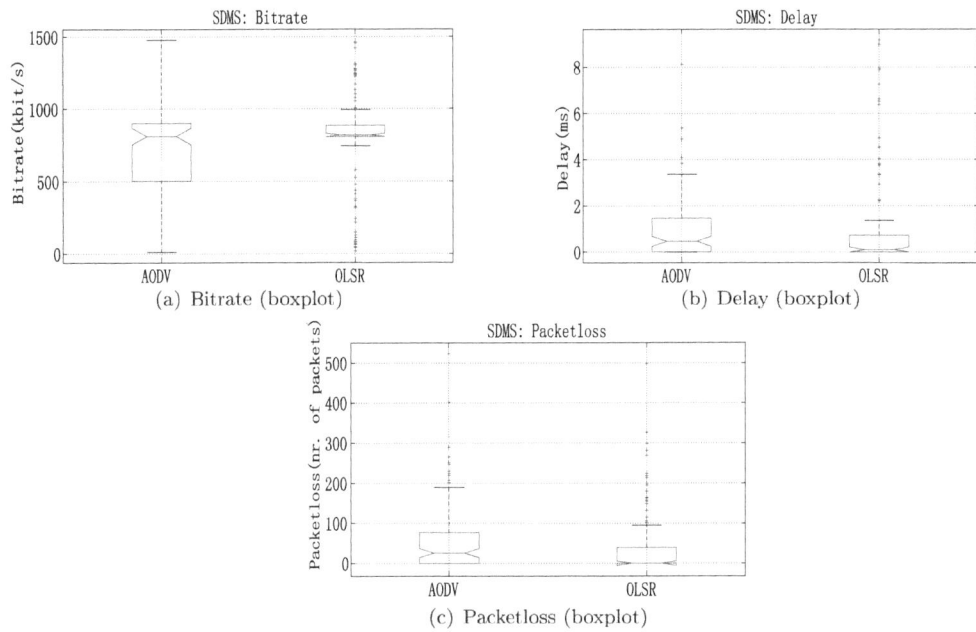

(a) Bitrate (boxplot)

(b) Delay (boxplot)

(c) Packetloss (boxplot)

Fig. 10. Different metrics vs different protocols for SDMS (boxplot).

## 6. Conclusions

In this paper, we used AODV and OLSR protocols for experimental evaluation and comparison and we implemented four scenarios (SS, SMS, DMS and SDMS) in a small MANET testbed of 7 nodes. We considered 3 metrics for performance evaluation: bitrate, delay and packetloss. We investigated the performance of MANET when two communicating nodes loose LOS during a period of time.

From our experimental results we found that, when the communicating nodes are moving and the routes change quickly, OLSR as a proactive protocol performs better than AODV, which is a reactive protocol.

In our future work, we would like to increase the number of nodes in our testbed and implement more realistic moving scenarios. We will run multiple flows between the communicating nodes and we will use the linear topology, in order to minimize the interference caused by multiple links in mesh topology.

## Acknowledgement

This work is support by a Grant-in-Aid for scientific research of Japanese Society for the Promotion of Science (JSPS). The authors would like to thank JSPS for the financial support.

## References

[1] A. Aikebaier, T. Enokido and M. Takizawa, Design and Evaluation of Reliable Data Transmission Protocol in Wireless Sensor Networks, *Mobile Information Systems* **4**(3) (2008), 237–252.

[2] L. Barolli, M. Ikeda, G. De Marco, A. Durresi and F. Xhafa, Performance Analysis of OLSR and BATMAN Protocols Considering Link Quality Parameter, *Proc of IEEE AINA-2009* (2009), 307–314.

[3] T. Clausen and P. Jacquet, Optimized Link State Routing Protocol, *Project Hipercom, INRIA Rocquencourt, France*, Technical Report, RFC 3626 (Experimental), (2003) Available: http://hipercom.inria.fr/olsr/draft-ietf-manet-olsr-11.txt.

[4] Core Software, AODV Software, Available on line at http://core.it.uu.se/core/index.php/ AODV-UU.

[5] A. Durresi, P. Zhang, M. Durresi and L. Barolli, Architecture for Mobile Heterogeneous Multi Domain Networks, *Mobile Information Systems* **6**(1) (2010), 49–63.

[6] D.S.J. De Couto, D. Aguayo, J. Bicket and R. Morris, A High Throughput Path Metric for Multi-hop Wireless Routing, *Proc of MobiCom-2003*, 9th Annual International Conference on Mobile Computing and Networking, (2003), 134–146, Available on line at http://dx.doi.org/10.1145/938985.939000.

[7] A. Dainotti, A. Botta and A. Pescap'e, Do You Know What You Are Generating?, *Proc of ACM CoNEXT-2007* (2007), 1–2.

[8] J. Goh and D. Taniar, Mining Frequency Pattern from Mobile Users, *Proc of the 8th International Conference on Knowledge-Based Intelligent Information and Engineering Systems KES-2004*, Part III, Lecture Notes in Computer Science, Springer, **3215** (2004), 795–801.

[9] J. Haemi, M. Fiore, F. Filali and C. Bonnet, A Realistic Mobility Simulator for Vehicular Ad Hoc Networks, *EURECOM Technical Report*, (2007), Available at: http://www.eurecom.fr/util/publidownload.en.htm?id=1811.

[10] A.M. Hanashi, I. Awan and M. Woodward, Performance Evaluation with Different Mobility Models for Dynamic Probabilistic Flooding in MANETs, *Mobile Information Systems* **5**(1) (2009), 65–80.

[11] M. Ikeda, L. Barolli, M. Hiyama, G. De Marco, T. Yang and A. Durresi, Performance Evaluation of Link Quality Extension in Multihop Wireless Mobile Ad-hoc Networks, *Proc of International Conference on Complex, Intelligent and Software Intensive Systems (CISIS-2009)* (2009), 311–318.

[12] W. Kiess and M. Mauve, A Survey on Real-world Implementations of Mobile Ad-hoc Networks, *Ad Hoc Networks* **5**(3) (2007), 324–339.

[13] V. Kawadia and P. Kumar, Experimental Investigations into TCP Performance over Wireless Multihop Networks, *Proc of E-WIND-2005, ACM SIGCOMM workshop on Experimental Approaches to Wireless Network Design and Analysis* (2005), 29–34.

[14] H. Lundgren, E. Nordstrom and C. Tschudin, Coping with Communication Gray Zones in IEEE 802.11b Based Ad hoc Networks, *Proc of WoWMoM-2002* (2002), 49–55.

[15]   P. Larbig, RaLink RT2570 USB Enhanced Driver, Available on line at http://homepages.tu-darmstadt.de/~p_larbig/wlan/.
[16]   Gj. Mino, L. Barolli, F. Xhafa, A. Durresi and A. Koyama, Implementation and Performance Evaluation of Two Fuzzy-based Handover Systems for Wireless Cellular Networks, *Mobile Information Systems* **5**(4) (2009), 339–361.
[17]   S.S. Manvi, M.S. Kakkasageri and J. Pitt, Multiagent Based Information Dissemination in Vehicular Ad Hoc Networks, *Mobile Information Systems* **5**(4) (2009), 363–389.
[18]   OLSR Download, Available on line at http://www.olsr.org/.
[19]   C. Perkins, E. Belding-Royer and S. Das, Ad hoc On-Demand Distance Vector (AODV) Routing, *RFC3561*, Nokia Research Center, University of California, University of Cincinnati, Technical Report, (2003) Available on line at http://www.ietf.org/rfc/rfc3561.txt.
[20]   M. Piorkowski, M. Raya, A.L. Lugo, M. Grossglauser and J.P. Hubaux, Joint Traffic and Network Simulator for VANETs, *Proc of Mobile and Information Communication Systems Conference* (*MICS-2006*), (October 2006) Available on line at: http://www.mics.ch/.
[21]   V. Pham, E. Larsen, O. Kure and P. Engelstad, Routing of Internal MANET Traffic over External Networks, *Mobile Information Systems* **5**(3) (2009), 291–311.
[22]   The Network Simulator NS-2, Available on line at http://www.isi.edu/nsnam/.
[23]   C. Tschudin, P. Gunningberg, H. Lundgren and E. Nordstrom, Lessons from Experimental MANET Research, *Ad Hoc Networks* **3**(2) (2005), 221–233.
[24]   D. Taniar and J. Goh, On Mining Movement Pattern from Mobile Users, *International Journal of Distributed Sensor Networks* **3**(1) (2007), 69–86.
[25]   K. Xuan, G. Zhao, D. Taniar and B. Srinivasan, Continuous Range Search Query Processing in Mobile Navigation, *Proc of IEEE ICPADS-2008* (2008), 361–368.
[26]   T. Yang, G. De Marco, M. Ikeda and L. Barolli, Impact of Radio Randomness on Performances of Lattice Wireless Sensors Networks Based on Event-reliability Concept, *Mobile Information Systems* **2**(4) (2006), 211–227.

**Elis Kulla** received his B.S and M.S degrees at Faculty of Information Technology, Polytechnic University of Tirana (PUT), Albania in 2007 and 2010, respectively. Presently, he is a Ph.D Student at Graduate School of Engineering, Fukuoka Institute of Technology (FIT), Japan. His research interests include ad-hoc networks, sensor networks, P2P networks, and vehicular networks.

**Masahiro Hiyama** received his B.S degree at Fukuoka Institute of Technology (FIT), Japan in 2009. Presently, he is a Master Student at Graduate School of Engineering, FIT, Japan. His research interests include ad-hoc networks and sensor networks.

**Makoto Ikeda** is an Assistant Research Fellow at the Center for Asian and Pacific Studies, Seikei University, Japan. He received B.S, M.S and Ph.D degrees in Information and Communication Engineering from Fukuoka Institute of Technology (FIT), Japan, in 2005, 2007, and 2010, respectively. He has published about 30 research papers in Journals and International Conference Proceedings. He won the Best Paper Award at NBiS-2008 International Conference. He has been a PC Member and Web Administrator for some International Conferences. His research interests include wireless networks, mobile computing, high-speed networks, P2P systems, ad-hoc networks and sensor networks. He is a member of the IEEE, ACM, IPSJ and IEICE.

**Leonard Barolli** received B.S and Ph.D degrees from Tirana University and Yamagata University in 1989 and 1997, respectively. From April 1997 to March 1999, he was a JSPS Post Doctor Fellow Researcher at Department of Electrical and Information Engineering, Yamagata University. From April 1999 to March 2002, he worked as a Research Associate at the Department of Public Policy and Social Studies, Yamagata University. From April 2002 to March 2003, he was an Assistant Professor at Department of Computer Science, Saitama Institute of Technology (SIT). From April 2003 to March 2005, he was an Associate Professor and presently is a Full Professor, at Department of Information and Communication Engineering, Fukuoka Institute of Technology (FIT). Dr. Barolli has published about 300 papers in referred Journals, Books and International Conference proceedings. He was an Editor of the IPSJ Journal and has served as a Guest Editor for many International Journals. Dr. Barolli has been a PC Member of many International Conferences and was the PC Chair of IEEE AINA-2004 and IEEE ICPADS-2005. He was General Co-Chair of IEEE AINA-2006, AINA-2008, AINA-2010, CISIS-2009 and CISIS-2010, Workshops Chair of iiWAS-2006/MoMM-2006 and iiWAS-2007/MoMM-2007, Workshop Co-Chair of ARES-2007, ARES-2008, IEEE AINA-2007 and ICPP-2009. Dr. Barolli is the Steering Committee Chair of CISIS and BWCCA International Conferences and is serving as Steering Committee Co-Chair of IEEE AINA, NBiS and 3PGCIC International Conferences. He is organizers of many International Workshops. Dr. Barolli has won many Awards for his scientific work and has received many research funds. He got the "Doctor Honoris Causa" Award from Polytechnic University of Tirana in 2009. His research interests include network traffic control, fuzzy control, genetic algorithms, agent-based systems, ad-hoc networks and sensor networks. He is a member of SOFT, IPSJ, and IEEE.

**Vladi Kolici** received his B.S and M.S degrees in Telecommunication Engineering from Polytechnic University of Tirana (PUT) in 1997 and 2005, respectively. He obtained his Ph.D from PUT in May 2009. From 1997 to 2004, he was a Research Associate and from 2005 to present he is a Lecturer at Department of Electronics and Telecommunications, Faculty of Information Technology, PUT. He is teaching several courses in the areas of wireless and mobile networking, P2P systems and quality of services. Dr. Kolici has published several papers in International and National Conference Proceedings in the areas of P2P and Ad-Hoc networks. Dr. Kolici received the Best Application Paper Award at the 6th International Conference on Advances in Mobile Computing and Multimedia (MoMM-2008) in 2008, Linz, Austria. His research interests include P2P networks, wireless and mobile networks, and high speed networks.

**Rozeta Miho** received her B.S and M.S degrees in Electronic and Telecommunication Engineering from Polytechnic University of Tirana (PUT), Albania, in 1985 and 1989, respectively. She obtained her Ph.D from PUT in November 1995. From 1985 to 1995, she was a lecturer, from 1996 to 2001 Assistant Professor, from 2001 to 2007 Associate Professor, and presently she is a Full Professor of PUT. From May 2009, she is the Dean of Faculty of Information Technology, PUT. She is teaching several courses in the areas of telecommunication networks, optical communications and optical fibre networks. Prof. Miho has published several papers in International and National Conference Proceedings in the areas of optical WDM networks, P2P and Ad-Hoc networks. She is the co-author of the Best Application Paper Award at the 6th International Conference on Advances in Mobile Computing and Multimedia (MoMM-2008) in 2008, Linz, Austria. Her research interests include P2P networks, optical networks, wireless networks and high speed networks.

# 4

# Recovery of flash memories for reliable mobile storages

Daesung Moon[a], Byungkwan Park[b], Yongwha Chung[c,*] and Jin-Won Park[d]

[a]*Knowledge Information Security Research Department, ETRI, 161 Gajeong-dong, Yuseong-gu, Daejeon, 305–700, Korea*

[b]*Department of Computer and Information Science, Sunmoon University, Asan, Chungnam 336–708, Korea*

[c]*Department of Computer and Information Science, Korea University, Jochiwon, Chungnam 339–700, Korea*

[d]*School of Games, Hongik University, Jochiwon, ChungNam 339–701, Korea*

**Abstract.** As the mobile appliance is applied to many ubiquitous services and the importance of the information stored in it is increased, the security issue to protect the information becomes one of the major concerns. However, most previous researches focused only on the communication security, not the storage security. Especially, a flash memory whose operational characteristics are different from those of HDD is used increasingly as a storage device for the mobile appliance because of its resistance to physical shock and lower power requirement. In this paper, we propose a flash memory management scheme targeted for guaranteeing the data integrity of the mobile storage. By maintaining the old data specified during the recovery window, we can recover the old data when the mobile appliance is attacked. Also, to reduce the storage requirement for the recovery, we restrict the number of versions to be copied, called Degree of Integrity (DoI). Especially, we consider both the reclaim efficiency and the wear leveling which is a unique characteristic of the flash memory. Based on the performance evaluation, we confirm that the proposed scheme can be acceptable to many applications as a flash memory management scheme for improving data integrity.

Keywords: Mobile storage, flash memory, data integrity

## 1. Introduction

As the value and the importance of the information stored in a storage device are increased proportionally with the storage capacity, there has been a growing interest in protecting the stored information against external attackers. Especially, with USB flash drives replacing floppy disks at the time of greater concern for security, sneakernet bandwidth has gone up considerably. Add to this mix the rise of mobile banking and other fiscal uses of mobile devices with embedded storage, and the risks increase dramatically [1].

However, *protecting information stored in a storage device* is a challenging problem because of the following reasons: third-party management of the corporate storage, widespread use of the mobile storage, and vulnerability of these devices to loss/theft/capture. Thus, *protecting information stored in mobile appliances* becomes an emerging topic in the storage system area. However, most previous

---

*Corresponding author. E-mail: ychungy@korea.ac.kr.

Fig. 1. The internal structure of a typical flash memory.

researches focused on either the mobile communication security issue [2–4] or the server-based storage security issue [5–8].

In this paper, we design and implement the mobile appliance storage for enhancing data integrity. Especially, we select a *flash memory* as our mobile appliance storage because of its resistance to physical shock and lower power requirement than Hard Disk Drive (HDD) [9–12]. We first propose the *Secure Flash Storage* by considering the characteristics of the flash memory and the requirements for data integrity, and then evaluate the performance of it with a trace-driven simulation.

The rest of the paper is structured as follows. Section 2 explains the overview of a flash memory and typical flash memory management schemes, and then describes the security issues in storage. Section 3 describes the proposed flash memory management scheme for improving data integrity, and the results of the performance evaluation are described in Section 4. Finally, conclusions are given in Section 5.

## 2. Background

A flash memory [12] is a type of Electrically Erasable Programmable Read Only Memory (EEPROM), and has the following characteristics over the HDD.

- Pros: small, light-weighted, robust; low power consumption; faster read access times.
- Cons: slower write access times; no in-place-update (needs an erase operation); limited lifetime ($<$ 100,000 times erasure); more expensive than disk (about 10 times); difficult to manage.

As shown in Fig. 1, the internal structure of a flash memory is defined by an *Erase Unit* (*EU*) which consists of 32$\sim$128 pages, and each page consists of a data area ($512/1024/2048B$) and a spare area ($16/32/64B$). In this paper, we consider $1Gbit$ NAND-type flash memories where the sizes of an EU, a page data area, a spare area are $16KB$ (32 pages), $512B$, and $16B$, respectively.

The basic operations defined over the flash memory are; "read a page" which takes less than $20usec$, "write (or program) a page" which takes less than $200usec$, and finally "erase an EU" which takes less than $2msec$. Note that, the basic unit of the read/write operation is a page whereas an EU is the basic unit of the erase operation. Because the erase operation is the most time consuming one, we need to be careful to manage it. The research topics related with the erase operation can be summarized as follows;

- Performance: reclaim (or garbage collection) efficiency.
- Wear Leveling: a limited number (about 100,000) of erases for each EU; need to erase EUs evenly.

Typical flash memory management schemes can be classified into two classes; Flash-Specific File Systems which are based-on "log-structured file system", such as JFFS [13], YAFFS [14]. The other one is *Flash Translation Layer* (*FTL*) [15] which emulates a block device using a flash memory and its upper layer interface is the standard file system, such as FAT [16]. We add some "recovery utility" to FTL in order to improve the data integrity of the flash memory.

The security issues in storage [17] can be summarized as follows;.

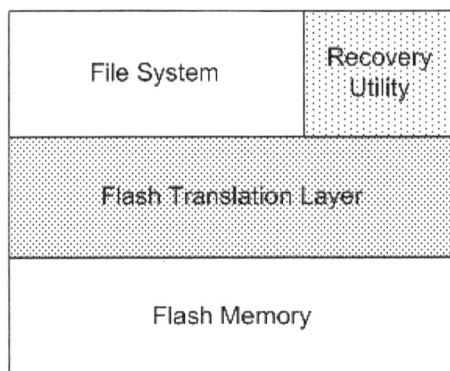

Fig. 2. Our Secure Flash Storage management scheme.

- Confidentiality: ensuring authorized users to access data, and can be achieved by encryption.
- Integrity: maintaining data consistency against accidental and malicious attacks to data through IDS or rollback.
- Performance: there is a tradeoff between the security level and the cost, and the most dominant cost is for encryption.

Several secure storage systems have been developed by considering the above security issues. For example, Network File System (NFS) [18] emphasizes the user authentication, Cryptographic File System (CFS) [19] provides the end-to-end security service by encrypting stored data, and Storage-based Intrusion Detection System (SIDS) [20] does not trust even the host machine operating system. Note that, all these approaches are "server-based" ones. To the best of our knowledge, the research result for securing "mobile storages" has not been reported yet.

## 3. Design of a secure flash storage

In this paper, we propose a new Flash Translation Layer (see Fig. 2) to improve the security of a mobile storage device for mobile appliances. That is, we apply some security concepts to the flash memory management scheme. Especially, we focus on the data integrity and propose some techniques such as roll-back after the attack, usage of timestamp, and page state managing. We assume that the data content can be protected by performing the user authentication first, rather than encrypting/decrypting repeatedly for all the data stored in the mobile storage. Of course, depending on the importance of the data, the data can be encrypted additionally. The issue of the user authentication is out of the scope of this research, but can be found in [21]. We also assume that the attack such as malware and denial-of-service is already detected using the security techniques [21]. Note that, regardless of whether we use the security techniques, we should have a recovery plan for incidents that will inevitably occur [22].

### 3.1. Data structures and state transition of page

For the purpose of explanation, we first describe the definition of the spare area for the NAND flash memory [23] in Fig. 3. Logical Sector Number (LSN) stores the information necessary for the mapping between the logical and the physical addresses. In this research, we utilize the reserved bytes (RSV) to design the Secure Flash Storage.

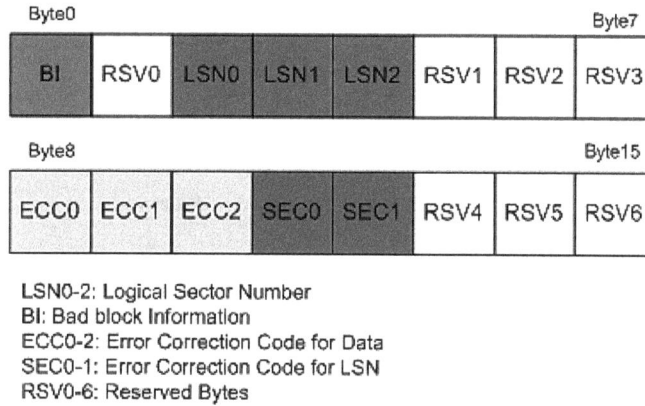

Fig. 3. Definition of the spare area for NAND flash memory [23].

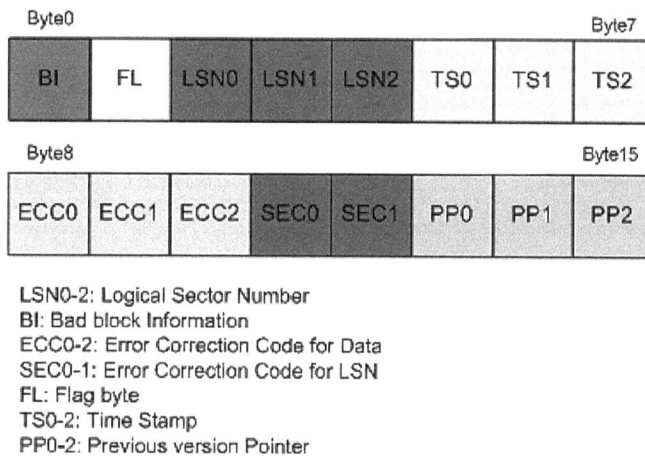

Fig. 4. New definition of the spare area for Secure Flash Storage.

Figure 4 shows the new definition of the spare area for our Secure Flash Storage. FL (Flag byte) represents the page state. In our design, we define a new state, called *Old* (*O*), in addition to *Free* (*F*), *Valid* (*V*), *Invalid* (*I*). Then, by maintaining the old data specified during the *Recovery Window*, we can recover the old data when the mobile appliance is attacked. TS (Time Stamp) in Fig. 4 represents the time for page write, and PP (Previous version Pointer) is for recovery. Especially, by following PP, we can reach the old data before the update.

According to the new definition of a state, we can represent the state transition of a page as Fig. 5. Note that, in order to save the extra space, the transition shown as "Update w/ Invalid" in Fig. 5 is defined for frequently updated data.

Then, we describe the necessary data structures for our Flash Translation Layer. *Direct Map* shown in Fig. 6(a) contains the logical-to-physical mapping information. It is built with Inverse Map in the mounting time and stored into RAM. On the contrary, *Inverse Map* shown in Fig. 6(b) contains the physical-to-logical mapping information in the spare area of the flash memory. The fields shown as bold ones are defined to implement our Secure Flash Translation Layer.

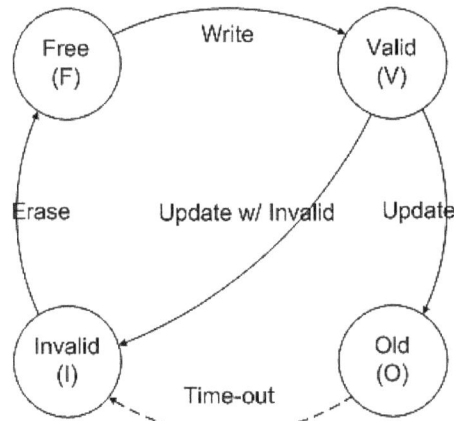

Fig. 5. State transition diagram of a page.

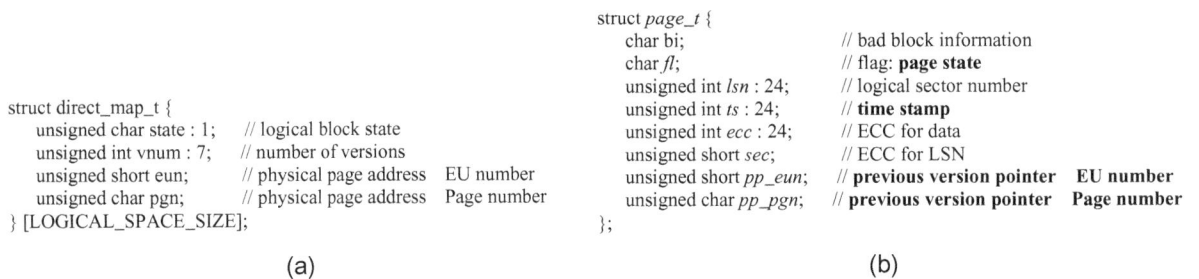

```
struct direct_map_t {
    unsigned char state : 1;    // logical block state
    unsigned int vnum : 7;      // number of versions
    unsigned short eun;         // physical page address   EU number
    unsigned char pgn;          // physical page address   Page number
} [LOGICAL_SPACE_SIZE];
```

(a)

```
struct page_t {
    char bi;                    // bad block information
    char fl;                    // flag: page state
    unsigned int lsn : 24;      // logical sector number
    unsigned int ts : 24;       // time stamp
    unsigned int ecc : 24;      // ECC for data
    unsigned short sec;         // ECC for LSN
    unsigned short pp_eun;      // previous version pointer   EU number
    unsigned char pp_pgn;       // previous version pointer   Page number
};
```

(b)

Fig. 6. (a) Direct map and (b) Inverse map.

## 3.2. Basic operations

We focus on the Page Update, the EU Reclaim, and the Space Thinning operations in this paper, although the Page Read and the Page Write operations are also the basic operations. And, the page allocation step in the Page Write and the Page Update is described in the EU Reclaim.

### 3.2.1. Page Read

To read a data, the logical block address needs to be translated to a corresponding physical block address (i.e., the EU number and the page number) through Direct Map stored in the main memory. Since the page read operation does not change any state information, it does not cause any update on Direct Map.

### 3.2.2. Page write

Page write operation writes a new value into an initially empty state (i.e., "E" in Direct Map), whereas page update operation writes a new value into an exist state (i.e., "X" in Direct Map). The page write procedure is as follows.

Step 1. Allocate new "F" state page.
Step 2. Write data on the page.

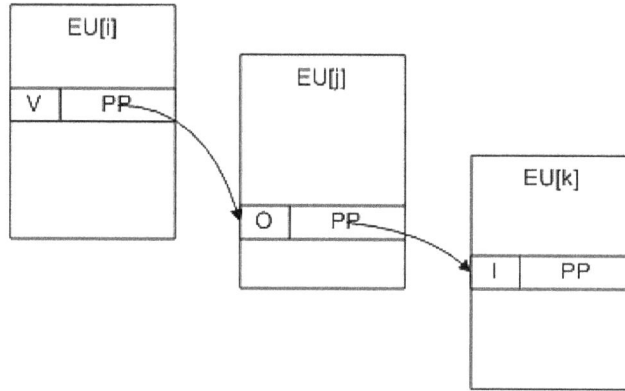

Fig. 7. Result of two page updates.

Step 3. Write management information on the spare area – change state "F" into "V" and timestamp. Since this operation write a new value into an empty state, previous version pointer (pp_eun, pp_pgn shown in Fig. 6(b)) is initialized with 1.

Step 4. Update the Direct Map (i.e., write the EU number and the page number, and change state "E" into "X").

### 3.2.3. Page update

The page update procedure is as follows, and the result of two updates is shown in Fig. 7. For example, after the first update, the states of the previous EU[k] and the current EU[j] become "O" and "V", respectively. After the second update, EU[k] may be changed as "I" state, which means the timestamp of EU[k] is beyond the Recovery Window.

Step 1. Allocate new "F" state page.

Step 2. Write data on the page.

Step 3. Write management information on the spare area – change state "F" into "V", timestamp, previous version pointer.

Step 4. Update previous version – change state "V" into "O".

Step 5. Update the Direct Map.

### 3.2.4. EU reclaim

With frequent updates, many "I" and "O" are created and "F" states are decreased. When we select an EU to reclaim, we should consider the reclaim efficiency and the wear leveling. In our scheme, we select an EU with the highest score computed in the following equation. In the following equation, we denote $valid$ $(j)$, $old$ $(j)$ and $invalid$ $(j)$ as the numbers of pages in EU $(j)$ with the corresponding states respectively. We set $\sigma(j)$ be the sum of the number of pages for recent versions of "O" states. Also, we set $\lambda$ be the parameter for wear leveling. That is, $\lambda$ approaches to 1 if there happens a problem with wear leveling and this case makes the second term large, resulting in selecting the EU. Otherwise, $\lambda$ approaches to 0 and we select the EU depending on the value of the first term in the following equation.

$$score(j) = (1 - \lambda)\left(\frac{valid(j) + invalid(j) + old(j)}{valid(j) + \sigma(j)}\right) + \lambda\left(\frac{\max_i\{erasures(i)\}}{1 + erasures(j)}\right) \quad (1)$$

where $0 < \lambda(\max_i\{erasures(i)\} - \min_i\{erasures(i)\}) < 1$.

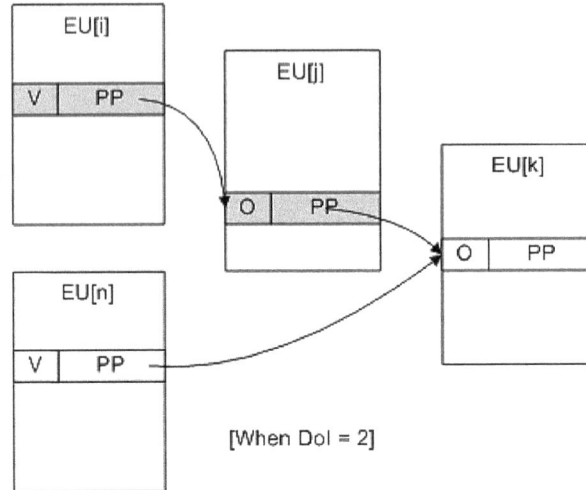

Fig. 8. Illustration of Space Thinning operation (DoI = 2).

```
struct eu_t {
            unsigned short erase_count;          // EU s erasure count
            float score;                         // score for reclaim
            struct pg_t {
                        unsigned char state;     // pages state
                        unsigned short pp_eun;   // previous pointer
                        unsigned char pp_pgn;    // previous pointer
                        unsigned int np_eun;     // next version pointer
                        unsigned char np_pgn;    // next version pointer
                        unsigned int ts ;        // timestamp
            } page[PG_NUM];
} EU [EU_NUM];
```

Fig. 9. EU state table stored in RAM.

### 3.2.5. Space Thinning

Increasing the Recovery Window size can enhance the data integrity with additional storage space. To balance the data integrity and the recovery space overhead, we introduce a notation of "*Degree of Integrity (DoI)*", which restricts the number of versions to be stored in order to make the versions to be evenly distributed in Recovery Window. DoI should be determined based on the characteristics of the applications and the size of the mobile storage device. The Space Thinning operation can be summarized as follows.

Step 1. When the number of versions reaches to $(2 \times DoI-1)$, start the following space thinning operation.

Step 2. Thinning out the even numbered versions from the tail of the list.

Step 3. The remained versions must be copied to new pages.

The illustration of the Space Thinning operation and the data structure required for reclaim and thinning are shown in Figs 8 and 9, respectively.

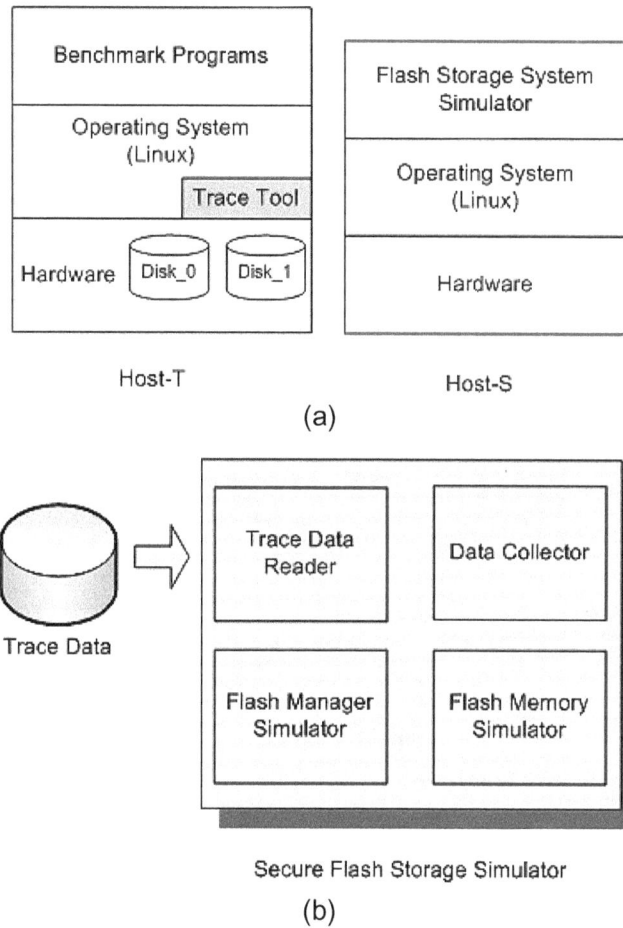

Fig. 10. (a) Experimental environment and (b) Structure of trace-driven simulator.

## 4. Performance evaluations

Since it is difficult to analyze the performance of the proposed Secure Flash Storage formally, we analyze it with a trace-driven simulation. Note that, most flash-based and/or disk-based previous researches also conducted a trace-driven simulation to verify their performance [24–26]. Especially, we want to verify that the caused overhead during normal operations is acceptable although it depends on DoI. The simulator for the Secure Flash Storage is driven by the trace data that were obtained by a patched Linux device driver to trace the device activities using the I/O benchmark program.

### 4.1. Simulation environment

We used one Linux-based server (Host-T) to obtain the trace data and another Linux-based server (Host-S) to develop and execute our simulator (see Fig. 10(a)). The trace tool for obtaining the trace data was implemented at the Linux device driver and we used the Disk Trace Tool developed at Bringham Young University [27]. Also, we used the benchmark program, called IOzone [28], to measure the I/O

```
struct trace_t {
        char type;              // access type:  r read,  w write
        char major;             // device s major number
        char minor;             // device s minor number
        char addr_type;         // L: LBA, C: CHS, S: SCSI
        unsigned int size;      // I/O request size
        unsigned int addr;      // IO Address
        unsigned int time;      // request time
};
```

Fig. 11. Trace data.

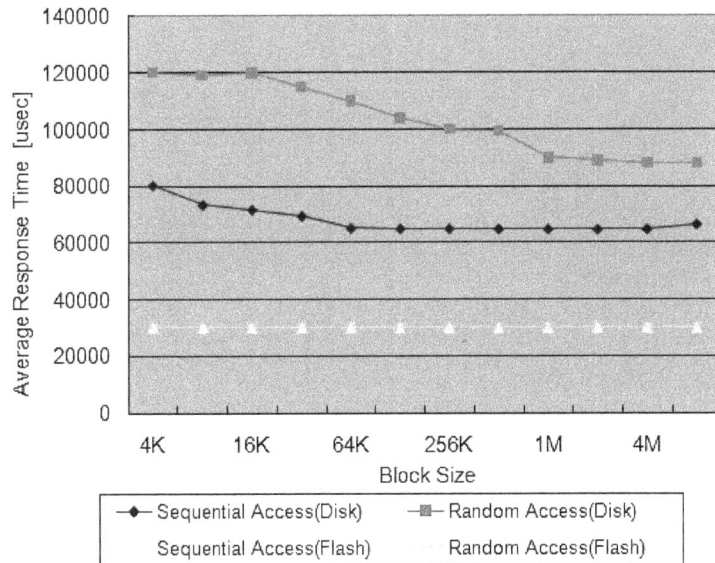

Fig. 12. Characteristics of flash memory and disk.

performance. Our simulator was implemented with C, and has the structure shown in Fig. 10(b). The data structure for the trace data is shown in Fig. 11.

### 4.2. Performance evaluation results

Figure 12 shows the difference of the average response times between the flash memory and the disk. Because the disk seek time is relatively large, the performance for sequential access and larger block is superior to that of random access and smaller block, respectively. On the contrary, the performance of the flash memory is almost independent of the access pattern (i.e., sequential vs. random) and the access size (i.e., large vs. small). The critical performance factor in the flash memory is the delay caused by the EU Reclaim. Also, the decision factor to the EU Reclaim execution is the *write bandwidth* which means the total data size to be written to the flash memory.

Figure 13 shows the characteristics of the Page Write operation. The 200 *usec* response time means the first write to a page, and the response time increases with the following page updates due to the required invalidate and write operations. If the write operations are executed increasingly, the "F" state pages decreases and the EU reclaim is needed. Once the EU Reclaim operation is initiated, however, the

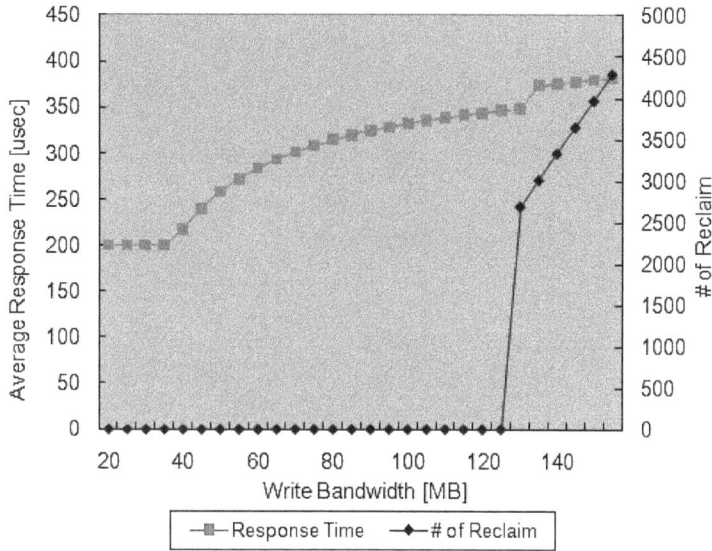

(a)

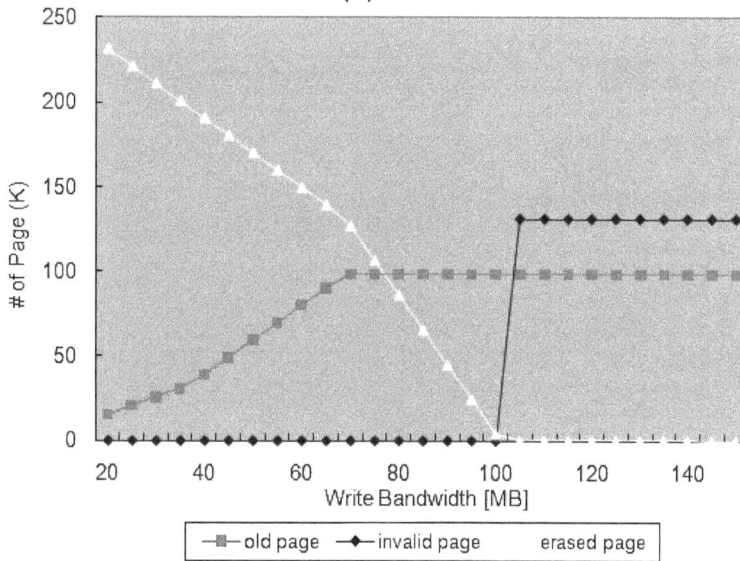

(b)

Fig. 13. Characteristics of page write (DoI = 3, sequential access pattern).

response time increases significantly. This is because the required time for the EU Reclaim is quite long (2*msec*) and the flash memory cannot do other operations during this EU Reclaim. The page state change in Fig. 11 also shows that maintaining "O" state pages increases the response time and requires more storage space. However, when we set DoI, the number of "O" state pages does not increase exponentially but approaches asymptotically to a certain level. This is due to the Space Thinning operation, and the number of "I" state pages increases. Also, when the number of "F" state pages is reduced to a certain level, the EU Reclaim operation is initiated.

In Fig. 14, the EU Reclaim operation causes the long response time when the write bandwidth reaches

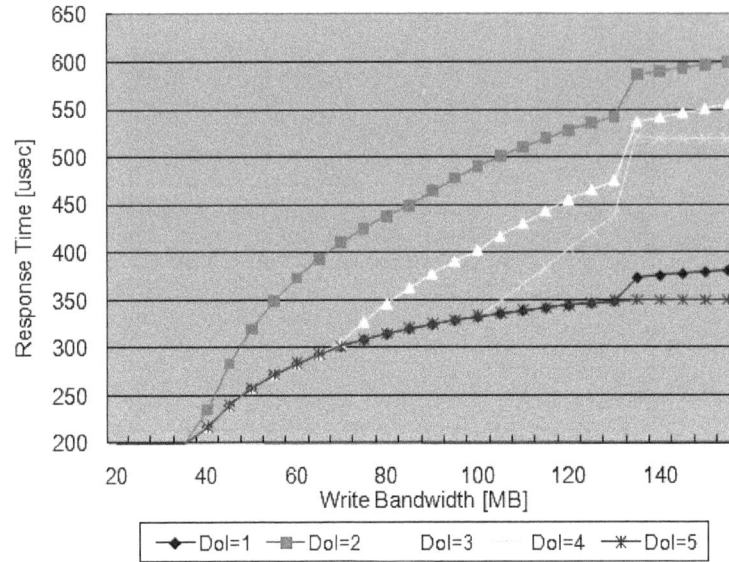

(a)

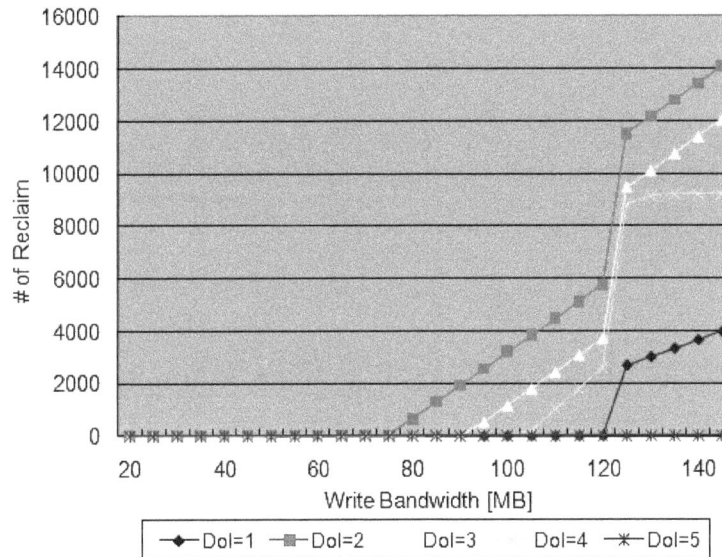

(b)

Fig. 14. (a) Page write time, (b) Reclaim frequency.

to 40*MB*/*sec* and the number of "F" state pages is reduced. DoI=2 needs more EU Reclaim operations than larger DoIs, and the jumps near 128*MB*/*sec* were from the target size (128*MB*) of our Secure Flash Storage.Also, the number of EU Reclaim operations is proportional to the write bandwidth.

Figure 15 shows the characteristics of the Space Thinning operation. When DoI is 2, the Space Thinning is initiated with lower write bandwidths and is increased proportionally with the write bandwidth. When we choose larger DoI, the first occurrence of the Space Thinning is delayed and the number of Space Thinning operations is reduced.

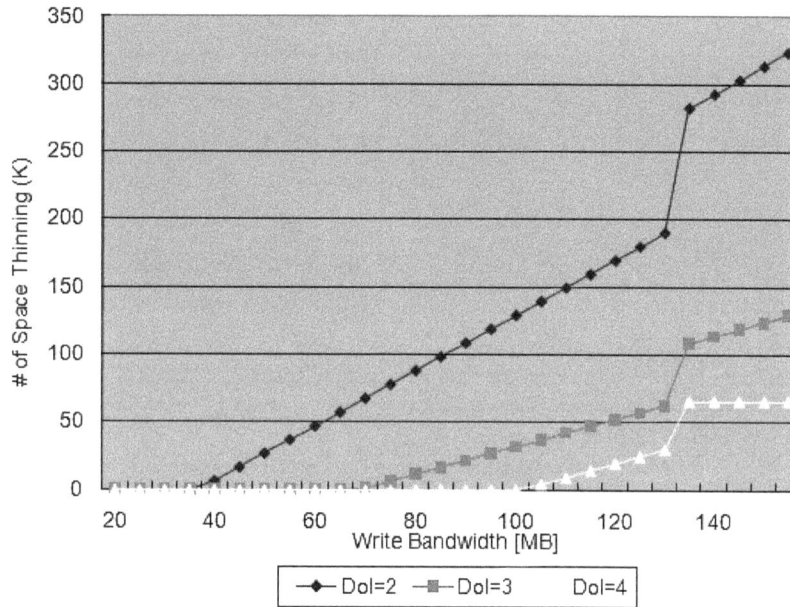

Fig. 15. Characteristics of Space Thinning.

Figure 16 shows the space overhead with different size of the Recovery Window (DoI = 10), with different trace data (DoI = 3), and with different DoI (recovery=20%). The Recovery Window shown as Fig. 16(a) was set as the relative value (i.e., 10%, 20%, ..., 100%) to the total trace time. The size of additional storage space is proportional to the increased size of the Recovery Window, but we know that we can control the size of the storage space depending on the application program and the DoI value. Also, the fact that the space overhead is not increased even with larger write bandwidths (see Fig. 16(c)) was due to the Space Thinning operation.

The results from the trace drvien simulation may not show the exact performance metrics for the Secure Flash Storage. However, the simulation results show the availability and superior performance of the Secure Flash Storage over hard disks for a typical workload.

## 5. Conclusions

As the mobile appliance is applied to many ubiquitous services and the importance of the information stored in it is increased, the security issue to protect the information needs to be considered. Although many researches reported the mobile communication security issue, the mobile storage security issue, especially with the flash memory whose operational characteristics are different from those of HDD, has not been reported.

In this paper, we proposed a flash memory management scheme, called *Secure Flash Storage*, targeted for guaranteeing the data integrity of the mobile storage. After defining some data structures and the necessary operations, two performance parameters (i.e., *Recovery Window*, *Degree of Integrity*) were derived to balance the security and the overhead by considering both the reclaim efficiency and the wear leveling. Finally, we implemented the Secure Flash Storage, and evaluated the Secure Flash Storage quantitatively with the performance parameters. Based on the trace-driven simulation, we confirm that

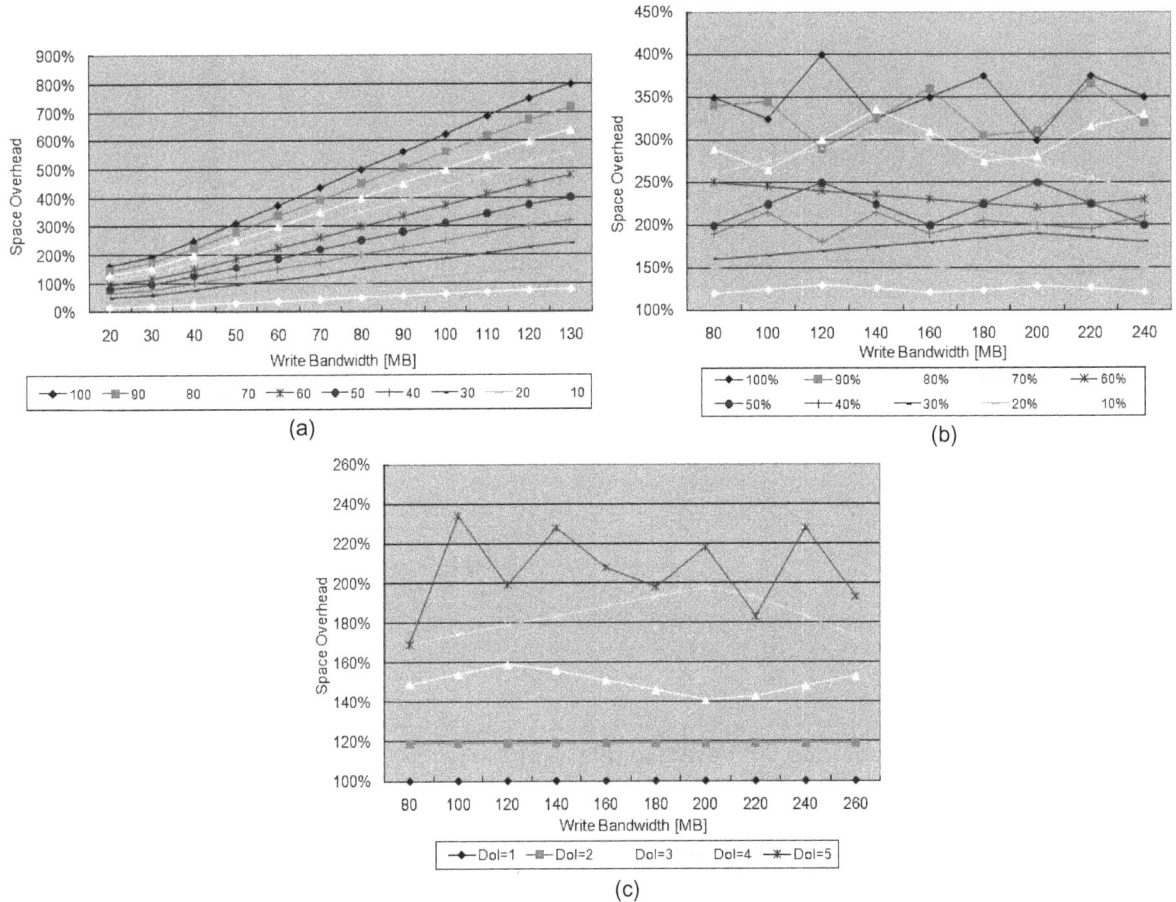

Fig. 16. Space overhead with different parameters; (a) Recovery window, (b) Trace data, (c) DoI.

the proposed scheme can improve the data integrity with an acceptable overhead by controlling the performance parameters.

## Acknowledgements

This research was partially financially supported by the Ministry of Education, Science Technology (MEST) and Korea Industrial Technology Foundation (KOTEF) through the Human Resource Training Project for Regional Innovation and partially by the MKE (The Ministry of Knowledge Economy), Korea, under the ITRC (Information Technology Research Center) support program supervised by the NIPA (National IT Industry Promotion Agency)

## References

[1]   J. Cole, Security in Storage: A Call for Participation, *IEEE Computer* **38**(9) (2005), 103–105.
[2]   A. Aikebaier, T. Enokido and M. Takizawa, Design and Evaluation of Reliable Data Transmission Protocol in Wireless Sensor Networks, *Mobile Information Systems* **4**(3) (2008), 225–237.

[3]   A. Durresi, M. Durresi and L. Barolli, Secure Authentication in Heterogeneous Wireless Networks, *Mobile Information Systems* **4**(2) (2008), 119–130.
[4]   D. Venugopal and G. Hu, Efficient Signature based Malware Detection on Mobile Devices, *Mobile Information Systems* **4**(1) (2008), 33–49.
[5]   *Proc. of Security in Storage Workshop*, IEEE, 2007.
[6]   *Proc. of Security in Storage Workshop*, IEEE, 2005.
[7]   *Proc. of Security in Storage Workshop*, IEEE, 2003.
[8]   *Proc. of Security in Storage Workshop*, IEEE, 2002.
[9]   G. Lawton, Improved Flash Memory Grows in Popularity, *IEEE Computer* **39**(1) (2006), 16–18.
[10]  S. Nath and P.B. Gibbons, Online maintenance of very large random samples on flash storage, *Proc. of the 34th conference on Very Large Data Bases (VLDB'08)* (2008), 970–983.
[11]  K. Park et al., Anticipatory I/O Management for Clustered Flash Translation Layer in NAND Flash Memory, *ETRI Journal* **30**(6) (2008), 790–798.
[12]  Samsung, Samsung Solid-State Disk Data Sheet, 2006.
[13]  D. Woodhouse, JFFS: The Journaling Flash File System, *Proc. of Ottawa Linux Symposium*, Available at http://sources.redhat.com/ jffs2/jffs2.pdf.
[14]  Aleph One, *YAFFS: Yet Another Flash Filing System*, Cambridge, UK, Available at http://www.aleph1.co.uk/yaffs/index.html, 2002.
[15]  A. Ban, *Flash File System*, US patent 5,404,485.
[16]  Silberschatz et al., *Operating System Concepts*, Wiley, 2003.
[17]  E. Riedel, M. Kallahalla and R. Swaminathan, A Framework for Evaluating Storage System Security, *Proc. of the 1st Conference on File and Storage Technologies*, 2002.
[18]  B. Pawlowski et al., *The NFS Version 4 Protocol*, SANE, 2000.
[19]  M. Blaze, A Cryptographic File System for UNIX, *Proc. of the 1st ACM Conference on Communications and Computing Security*, 1993.
[20]  A. Pennington et al., Storage-based Intrusion Detection: Watching Storage Activity for Suspicious Behavior, *Proc. of the 12th USENIX Security Symposium*, 2003.
[21]  W. Stallings, *Cryptography and Network Security: Principles and Practice*, Prentice Hall, 2006.
[22]  S. Farrell, Portable Storage and Data Loss, *IEEE Internet Computing* **12**(3) (2008), 90–93.
[23]  Samsung, *NAND Flash Spare Area Assignment Standard*, 2005.
[24]  L. Chang and T. Kuo, Real-time Garbage Collection for Flash Memory Storage Systems of Real-Time Embedded Systems, *ACM Transactions on Embedded Computing Systems* **3** (2004), 837–863.
[25]  C. Park et al., Cost-Efficient Memory Architecture Design of NAND Flash Memory Embedded Systems, *Proceedings of the 21st International Conference on Computer Design (ICCD'03)* (2003), 474–480.
[26]  J. HSIEH, T. KUO and L. CHANG, Efficient Identification of Hot Data for Flash Memory Storage Systems, *ACM Transactions on Storage* **2**(1) (2006), 22–40.
[27]  F. Sorenson et al., A System-Assisted Disk I/O Simulation Technique, *Proc. of the 7th International Symposium on Modeling, Analysis and Simulation of Computer Telecommunication Systems* (1999), 296–304.
[28]  D. William and D. Capps, *IOzone File System Benchmark*, http://www.iozone.org/, 2006.

---

**Daesung Moon** received the MS degree from Busan National University, Korea, in 2001. He received the PhD degree from the Korea University, Korea in 2007. He joined the Electronics and Telecommunications Research Institute (ETRI), Korea, in 2000, where he is currently a Senior Member of the engineering staff in the Human Recognition Technology Research Team. His research areas are biometrics, image processing, and security.

**Yongwha Chung** received the BS and MS degrees from Hanyang University, Korea, in 1984 and 1986. He received the PhD degree from the University of Southern California, USA in 1997. He worked for ETRI from 1986 to 2003 as a Team Leader. Currently, he is an Associate Professor in the Department of Computer Information, Korea University. His research interests include biometrics, security, and performance optimization.

**Byungkwan Park** received the BS degree in Electronic Engineering from Hanyang University, and MS degree in Computer Science from Korea Advanced Institute of Science and Technology. He received the PhD degree in Computer Science from Korea University. He is currently a professor in Division of Computer Science and Engineering at Sunmoon University. He previously worked as senior engineer in ETRI. His research interests include computer architecture, embedded systems and storage technologies.

**Jin-Won Park** graduated from Seoul National University in Korea. He received PhD degree from The Ohio State University in USA in 1987, majoring in industrial and systems engineering. He had been working at Electronics and Telecommunication Research Institute(ETRI) in Korea (1988–1999). He is currently teaching at Hongik University in Korea. His research interest is in the area of the performance evaluation of computer systems.

# 5

# A protocol for content-based communication in disconnected mobile ad hoc networks

Julien Haillot and Frédéric Guidec*
*Laboratoire Valoria, Université de Bretagne Sud / Université Européenne de Bretagne, Rennes, France*

**Abstract.** In content-based communication, information flows towards interested hosts rather than towards specifically set destinations. This new style of communication perfectly fits the needs of applications dedicated to information sharing, news distribution, service advertisement and discovery, etc. In this paper we address the problem of supporting content-based communication in partially or intermittently connected mobile ad hoc networks (MANETs). The protocol we designed leverages on the concepts of opportunistic networking and delay-tolerant networking in order to account for the absence of end-to-end connectivity in disconnected MANETs. The paper provides an overview of the protocol, as well as simulation results that show how this protocol can perform in realistic conditions.

Keywords: Mobile ad hoc network, delay/disruption tolerant networking, opportunistic networking, content-based networking

## 1. Introduction

Applications dedicated to information sharing, news distribution, or service advertisement and discovery, all share a common characteristic: they require a communication model where information can flow towards any interested receiver rather than towards set destinations.

Content-based communication is a style of communication that fits perfectly the needs of such applications. In content-based communication, the flow of information is interest-driven rather than destination-driven [5]. Receivers specify the kind of information they are interested in, without regard to any specific source. Senders simply send information in the network without addressing it to any specific destination.

In this paper we address the problem of supporting content-based communication in a disconnected mobile ad hoc network (MANET). A MANET can become disconnected when, for example, the mobile hosts that compose the network are very sparsely or irregularly distributed. The whole network then appears as a collection of distinct "islands". Communication between hosts that belong to the same island is possible, but no temporaneous communication is possible between hosts that reside on distinct islands. Figure 1 shows a disconnected MANET, which is typical of the kind of network we consider in our work. This MANET is composed of a number of laptops carried by users, which can move in and between buildings (for example, the buildings of a campus). In this example, some laptops are temporarily isolated (either because there is no other laptop within their transmission range, or more simply because they have been put in suspend mode for a while), while other laptops have a number of neighbours, with which they can try to communicate using either single-hop or multi-hop transmissions.

---

*Corresponding author. E-mail: Frederic.Guidec@univ-ubs.fr.

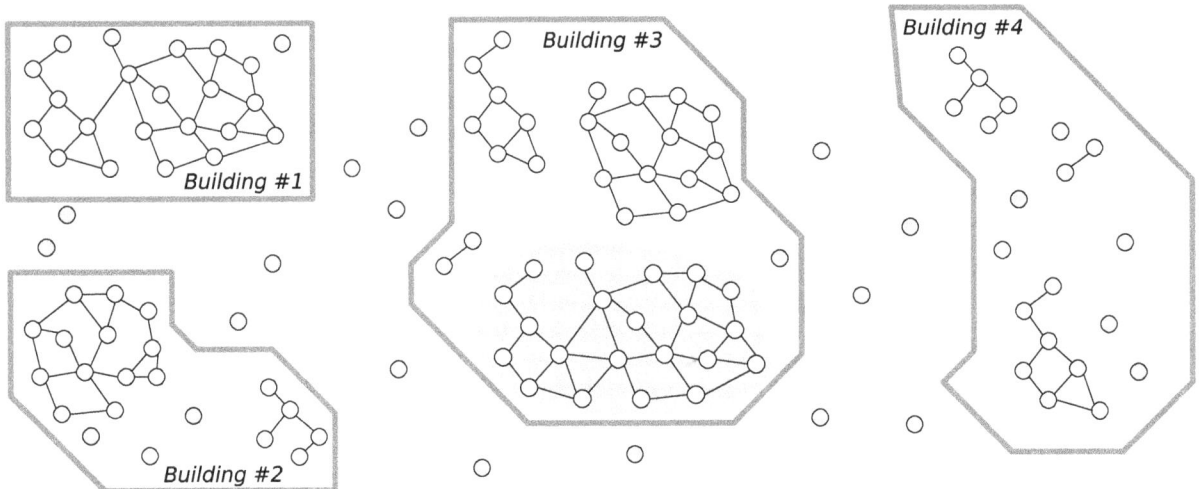

Fig. 1. Example of a disconnected MANET.

In fully connected wired networks, content-based communication can be achieved by constructing an underlying communication system, whose role is to forward each piece of information from its sender to all interested hosts [5]. This system is usually organised as a logical, content-driven routing infrastructure, which itself can be implemented as an overlay network that covers the whole physical point-to-point network. This approach can hardly be applied in a disconnected MANET, since in such a network the absence of end-to-end connectivity between distinct islands precludes building any network-wide overlay.

In such conditions, a method must be devised in order to bridge the gap between non-connected parts of the network. Delay-tolerant networking is an approach that can help with that respect [12]. In a delay-tolerant network, a message can be *stored* temporarily on a host, in order to be *forwarded* later by this host when circumstances permit. If the network includes mobile hosts – or if all hosts are mobile – then mobility becomes an advantage, as it makes it possible for messages to propagate network-wide, using mobile hosts as *carriers* that can move between remote – and possibly non-connected – fragments of the network. In a disconnected MANET such as that shown in Fig. 1, people moving between buildings (or between different parts of a building) can thus contribute to disseminate information between non-connected fragments of the MANET. Figure 2 shows a typical example, where the laptop of a user moving between two groups of users can contribute to carry information between these two groups.

In the remainder of this paper we present the main features of a middleware platform we designed along these lines. An overview of this platform is given in Section 2, and details about this platform's components are provided in later sections. The way information differentiation is performed in this platform is notably described in Section 3. The platform actually supports the opportunistic, content-driven dissemination in a disconnected MANET of structured pieces of information we refer to as *documents* (Section 3.1). Application services running on a mobile host can subscribe for receiving particular kinds of documents by specifying a *pattern* characterizing each category of desired documents. The patterns defined by all the subscribers running on the same host are combined and constitute the host's *interest profile* (Section 3.2). The dissemination of documents in the network relies on a *gossip-based protocol*: transient contacts between mobile hosts are exploited to exchange documents between these hosts, according to their respective interest profiles (Section 4). A host that subcribes for receiving a particular kind of documents is expected to serve as a mobile carrier for these documents. However

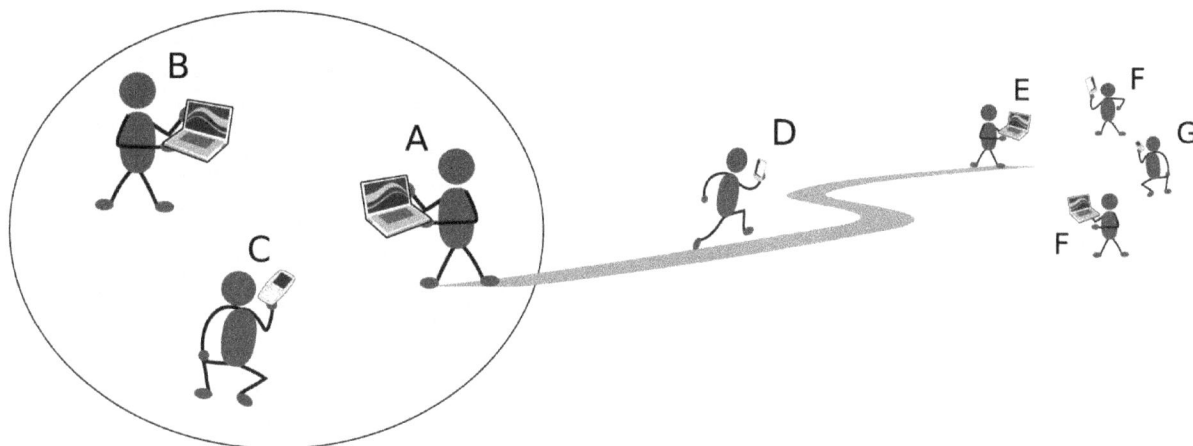

Fig. 2. Detail of Fig. 1, showing how users moving in the campus can carry information between connectivity islands.

it can also serve as an altruistic carrier for documents it is not especially interested in, provided this behaviour does not compromise its chances of collecting and carrying interesting documents.

The protocol implemented in the platform is actually defined as a two-layer stack. The *upper layer* (Section 4.1) defines how neighbour hosts can discover each other and exchange documents. It also provides means for storing documents in a local cache, so each host can serve as a mobile carrier while roaming the network. The *lower layer* of the protocol (Section 4.2) makes it possible for hosts that reside in a connected fragment of the network – such as the greyish area in Fig. 1 – to interact with n-hop neighbours using temporaneous multi-hop forwarding. This approach helps disseminate documents faster and more efficiently in the network.

The whole protocol was designed so as to be very frugal regarding the resources it consumes, and yet efficient at disseminating documents. Simulation results presented in Section 5 confirm this claim, and show how our platform can perform for disseminating documents in an environment such as that shown in Fig. 1.

Related works – and especially works we took inspiration from – are discussed in Section 6. In Section 7 we conclude this paper, listing a number of directions we contemplate investigating in future work.

## 2. Overview of the system

Figure 3 provides an overview of DoDWAN (stands for Document Dissemination in disconnected Wireless Ad hoc Networks), the middleware platform we designed in order to support content-based information dissemination in disconnected mobile ad hoc networks. This platform is not just simulation code: it has been fully implemented in Java, and is now being used for developing effective application-level software.

DoDWAN is distributed under the terms of the GNU General Public License.[1] It provides high-level application services with means to publish and subscribe for structured pieces of information we refer to as *documents*, using a dedicated API. Each host is expected to allocate a certain amount of storage

---

[1] http://www-valoria.univ-ubs.fr/CASA/DoDWAN.

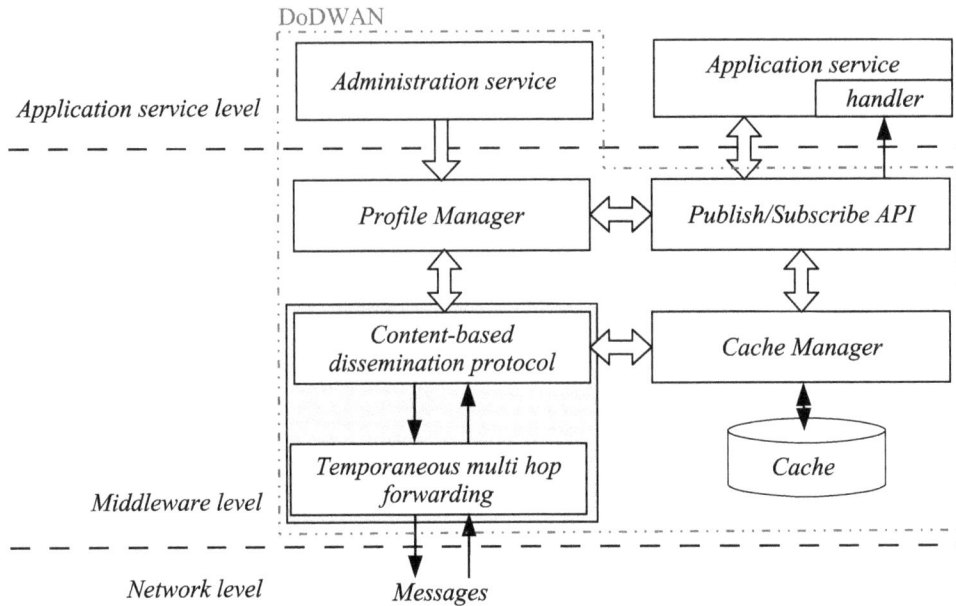

Fig. 3. Architecture of the communication platform.

space for implementing a local cache, in which documents can be stored for a while. Although we make no assumption about the storage capacity on each mobile host, it is assumed that this capacity is bounded, and that it can actually be different on different hosts. Each host's cache is therefore under the responsability of a local cache manager that can decide what documents can be put in the cache, and that can remove them from the cache whenever needed. Note that in this paper our objective is not to evaluate the merits of different strategies for cache management, as this kind of work has been done (e.g. in [14]).

When a document is published by a local application service through the Pub/Sub API, it is simply put in the host's local cache. Afterwards this document can disseminate from host to host in the network, using as carriers and forwarders those hosts that are willing to help in this dissemination.

Conversely, in order to subscribe for a particular kind of document, an application service must use the Pub/Sub API to specify a pattern that characterizes this kind of document. This pattern is then passed to a profile manager, that keeps track of all local subscription patterns and combines them to define the host's so-called "interest profile" (which somehow characterizes the whole set of documents the host is interested in). This interest profile can also be directly altered by a dedicated administration service. Thus, the user or administrator of a mobile host can if needed bypass the standard Pub/Sub API in order to indicate that this host should serve as a mobile carrier for specific categories of documents, even if these documents are of no direct interest to local application services.

The two remaining blocks (that are highlighted by a grey background in Fig. 3) are the key elements to content-based networking in our system. They implement a two-layer protocol we designed to support opportunistic, content-driven interactions between mobile hosts. This protocol can be perceived as a particular implementation of the – somewhat abstract – Autonomous Gossipping (A/G) algorithm described in [10]. It allows mobile hosts to exploit transient contacts for exchanging information according to their respective interest profiles. Interaction between mobile hosts relies on a simple model, whereby each host periodically informs other hosts located in its neighbourhood about its own interest profile and about the documents that are currently available in its local cache. When a host discovers that one of its

```
<descriptor>
  ...
</descriptor>
<payload>
  ...
</payload>
```

Fig. 4. General structure of a document.

neighbours can provide a document it is interested in (that is, a document that matches its own interest profile and that is not already available in its own cache), it can request a copy of this document from this neighbour. Transient contacts between mobile hosts are thus exploited opportunistically for exchanging documents between these hosts, based on their respective interest profiles, and based on the documents they can provide each other on demand.

In the remainder of this paper we provide a detailed description of this two-layer protocol we implemented in the DoDWAN platform, motivating the role of each layer, and showing how they together support content-driven gossiping between mobile hosts. In Section 3 we specify how content-based information differentiation is realized in our system and in Section 4 we detail how content-based information dissemination is carried out in our protocol.

## 3. Information differentiation: The key to content-driven dissemination

Keys to content-driven information dissemination are:

1. the ability to differentiate pieces of information based on their content
2. the ability to specify the kinds of information each subscriber is interested in.

In our system, the "pieces of information" we consider are actually referred to as "documents". Selection (or filtering) patterns are used to describe what kinds of documents a subscriber is interested in, and interest profiles aggregate the patterns of all subscribers running on a single host, and therefore define the whole set of documents this host is interested in.

### 3.1. Documents, descriptors and identifiers

A document is actually composed of two parts (see Fig. 4): its descriptor, and its content (or payload). In this paper we use an XML dialect to illustrate most of the data structures we deal with, including messages and message parts. This is actually for the sake of clarity only. In practice, the system we designed uses more effective (i.e. binary) formats for data structures and messages.

The descriptor can be perceived as a collection of attributes, which can provide any kind of information about the corresponding document, such as its identifier, its origin, its topic, a list of keywords, the type of its content, etc. In our system the identifier is the only mandatory attribute in a descriptor. Each document must be assigned a unique identifier value, as this value is used as a means to differentiate documents and to identify duplicate copies of a document. Besides, as a general rule we assume that the size of a document is far greater than that of its descriptor, which is itself significantly greater than that of its identifier (typical orders of magnitude are $O(10\ kB)$ for a document, $O(100\ B)$ for its descriptor, and

```
<descriptor>    // (D1)
  id="254d3g64z36cd"
  service="filesharing"
  type="application/pdf"
  date="Fri Jul 11 09:52:11 CEST 2008"
  deadline="Sat Jul 12 14:00:00 CEST 2008"
  publisher="Fred"
  keywords="mobile,ad hoc,delay-tolerant,\
            opportunistic"
</descriptor>
```

```
<descriptor>    // (D2)
  id="3ab7285ef6548"
  service="news"
  group="comp.networking"
  type="text/rfc850"
  date="Thu Jul 10 10:52:11 CEST 2008"
  publisher="Julien"
  keywords="mobile,ad hoc"
</descriptor>
```

Fig. 5. Examples of document descriptors.

$O(10\ B)$ for its identifier). Our protocol leverages on this contrast between the size of a document and that of its descriptor or identifier whenever possible in order to minimise the amounts of data exchanged by neighbouring hosts.

Examples of document descriptors are shown in Fig. 5. The first descriptor, labelled $D1$ in the figure, is that of a document published within the context of a (hypothetical) filesharing service. It specifies what is the type of the document (in that case it is a PDF document), and it includes keywords characterizing this document. Note that the identity of the publisher is specified in the descriptor (although that is not a requirement), as well as indications about when the document was published and when it should be considered as being obsolete.

The second descriptor ($D2$) is that of a document published within the context of a peer-to-peer newsgroup-based discussion service. In that case the newsgroup the document is published in is specified by attribute *group*, and the document's payload type is plain text with a standard RFC 850 (NNTP) header.

According to the principle of content-based networking, any piece of information contained in a document's descriptor can bring useful indication about how this document should be managed by a mobile host. By matching a document's descriptor against its own interest profile, a host can decide if this document should be put in its local cache. Yet indications provided in the descriptor can also help the local cache manager decide what documents should be removed from the cache, and arbitrate between conflicting documents. For example descriptor $D1$ specifies explicitly the deadline of the corresponding document, whereas descriptor $D2$ contains no such information. The information provided in $D1$ is thus a clear indication that any host that maintains a copy of the document described by $D1$ in its local cache can discard this copy after the set deadline. In contrast a host that carries a copy of the document described by $D2$ must decide freely when this copy should be removed from its cache.

```
<pattern>     // (P1)
  service="filesharing"
</pattern>
```

```
<pattern>     // (P2)
  service="filesharing"
  type="application/(gif|jpg|png)"
</pattern>
```

Fig. 6. Examples of document descriptors.

Document descriptors can also include attributes specifying priority levels, so as to help mobile hosts decide which documents they should preferably maintain in their local cache. Indeed, in this work our objective is to design a system such that, whenever a host is proposed a new document by another host, it is able to decide whether it is worth receiving and storing this document, considering what is already present in its local cache. Again document differentiation is the key to this approach, as it makes it possible for each host to classify documents based on their descriptors, and decide that some documents are actually more important than others.

### 3.2. Selection patterns and interest profiles

#### 3.2.1. Selection patterns

Document differentiation is performed based on the information available in document descriptors, and this differentiation relies on selection patterns. For practical reasons, we define a selection pattern as a collection of attributes whose values take the form of regular expressions. A pattern is said to *match* a document's descriptor if, for each attribute defined in the pattern, the same attribute exists in the descriptor, and if the value of the descriptor's attribute matches the regular expression of the pattern's attribute. Figure 6 shows two simple patterns. Let us examine how these patterns patterns in Fig. 6 match against the descriptors in Fig. 5. Obviously pattern *P1* matches descriptor *D1*, since attribute *service* is defined and has exactly the same value in *P1* and in *D1*. In contrast *P1* does not match descriptor *D2*, for the value of attribute service in *D2* does not match that defined in *P1*. Pattern *P2* does not match descriptor *D1*. Although *P2* and *D1* both carry the same *service* attribute, the regular expression defined for attribute *type* in *P2* does not match the value of this attribute in *D1* (*P2* actually allows to select only documents that contain images in either JPEG, GIF, or PNG format, whereas *D1* is the descriptor of a PDF document). Finally *P2* does not match *D2*, since the values of the *service* attribute differ in both structures.

#### 3.2.2. Interest profiles

The interest profile of a host determines the different kinds of documents it is interested in, and thus the kinds of documents for which it is willing to serve as a mobile carrier. It is defined as an aggregate of all selection patterns defined by local subscribers (plus possibly additional patterns defined through the platform's administration service). The interest profile is therefore notably updated whenever a local application service subscribes for a new kind of documents (or cancels a former subscription) through the platform's Pub/Sub API.

```
<profile>      // (host B)
  <pattern>
    service=''news''
    group=''comp.networking|alt.fan.science-fiction''
  </pattern>
  <pattern>
    service=''filesharing''
    keywords=''mobile|ad hoc''
  </pattern>
</profile>
```

Fig. 7. Interest profile of host *B*.

```
<profile>      // (host C)
  <pattern>
    publisher=''Fred''
  </pattern>
</profile>
```

Fig. 8. Interest profile of host *C*.

Whenever the host is offered a document by one of its neighbours, it must decide whether this document is an interesting one or not. A document's descriptor is said to *match* a host's profile if it matches at least one of the patterns defined in this profile.

Examples of possible interest profiles are shown in Figs 7 and 8. Let us respectively call *B* and C the hosts that present these two profiles. In that case *B*'s profile consists of two basic selection patterns, which indicate that *B* is interested:

- in documents – or articles, for that matter – published within the context of the newsgroup service and pertaining to any of the two specified newsgroups;
- in documents published within the context of the filesharing service and characterized by any combination of the specified keywords.

Likewise, *C*'s profile indicates that it is interested in any document that has been published by user Fred.

## 4. Communication protocol

As explained in Section 2 the gossip-based protocol implemented in our platform is actually defined as a two-layer stack. The upper layer supports the content-driven, delay-tolerant dissemination of documents in the network. It notably provides support for storing documents in a host's local cache, so this host can serve as a mobile carrier for these documents while moving in the network. It also defines how neighbour hosts can interact in order to exchange documents according to their respective interest profiles.

Neighbour hosts are hosts that temporarily reside on the same connected fragment of the network. Interaction between such hosts requires that they be able to communicate using either single-hop or multi-hop transmissions depending on their location in the network. The lower layer of the protocol provides mechanisms for temporaneous, multi-hop forwarding, which is required in the latter case.

*Frugal use of the wireless medium.* While designing this protocol, we strived to make it as frugal as possible regarding the resources it consumes, and especially regarding its consumption of wireless bandwidth. Both the number and the size of the messages required for disseminating documents are systematically kept at a minimum.

Moreover, in this work we assume that mobile hosts communicate using the Wi-Fi technology, which supports two distinct transmission modes, depending on whether data frames are sent in unicast mode or in broadcast mode. We therefore chosed to rely (as suggested in [1,30]) on broadcast transmissions rather than on unicast transmissions whenever possible. This point actually needs further explanation, as it is not common to favour broadcast transmission over unicast transmission when dealing with wireless ad hoc communication.

With the Wi-Fi (a.k.a. IEEE 802.11) technology, broadcast transmissions are admittedly less reliable than unicast transmissions. This is mostly because an ARQ (Automatic Repeat-Query) mechanism is implemented at MAC level, and this mechanism is used only when a frame is sent in unicast mode. Sending a unicast data frame is therefore a fairly – though not totally – reliable operation, as the sender keeps re-sending the frame until it receives an ACK frame from the destination host, or until the maximum number of retransmissions has been reached. In contrast sending a data frame in broadcast mode implies no such mechanism. The frame is sent once and once only on the medium. If interferences occur during this transmission, the frame can be lost without the sender and/or receiver(s) being aware of this loss. Sending a data frame in broadcast mode is thus more risky than sending it in unicast mode, but it is also significantly less costly, since it implies no ACK frames, and no retransmission. It also makes it possible to send a message to all direct neighbours of the sender using a single broadcast frame, rather than by using a round of unicast frames addressed to each neighbour successively.

While designing our protocol we decided to favour a frugal consumption of resources, to the detriment of the reliability of transmissions. We therefore use broadcast transmissions whenever possible, while ensuring that the protocol is totally resilient to transmission failures. As a general rule, interactions between neighbour hosts rely on an opportunistic exchange scheme rather than on a strict transactional scheme. Each host only maintains soft-state information about its neighbourhood, and no communication session is ever established between neighbours.

## 4.1. Support for content-driven, delay-tolerant document dissemination (upper layer)

In our system, interaction between mobile hosts relies on a simple model, whereby each host periodically informs other hosts located in its neighbourhood about its own interest profile and about the documents that are currently available in its local cache. When a host discovers that one of its neighbours can provide a document it is interested in (that is, a document that matches its own interest profile and that is not already available in its own cache), it can request a copy of this document from this neighbour. Transient contacts between mobile hosts are thus exploited opportunistically for exchanging documents between these hosts, based on their respective interest profiles, and based on the documents they can provide each other on demand.

The system is mostly event-based: each host simply reacts to internal events (such as a timer triggering periodic tasks) and external events (such as the receipt of a message). The protocol is presented in pseudo-code in Section 7. Details about the main parts of this code are provided below.

### 4.1.1. Announcing one's catalog and personal interest profile
Each host $n_i$ periodically broadcasts an announce that combines all or part of the following elements:

```
<announce>      // (Comprehensive form)
  from="host_id"
  key="24f6g4dq6"
  <profile>
    // As shown in Fig. 7 and 8
  </profile>
  <catalog>
    // As shown in Fig. 10
  </catalog>
</announce>
```

```
<announce>      // (Short form)
  from="host_id"
  key="24f6g4dq6"
</announce>
```

Fig. 9. Examples of the two forms of announces (one complete, one short) a host can broadcast periodically.

- the host's identity $n_i$
- a hash-key $k$
- a description of its own interest profile $prof(n_i)$
- a catalog $cat(n_i)$, which contains the descriptors of locally cached documents that can be of interest to its neighbours

This announce is broadcast as a single control message, whose propagation scope can be set explicitly by the sender (this is explained further in Section 4.2).

An example of an announce is shown in Fig. 9. Note that each announce can actually be broadcast either in a comprehensive (meaning long) form, or in a short form. Let us first consider the elements contained in a comprehensive announce.

By broadcasting an announce that contains its identity, a host allows its neighbours to discover or confirm that it is itself one of their current neighbours. By also inserting its own interest profile in this announce, it lets them know what kinds of documents it is interested in. Conversely, by receiving similar announces from its neighbours, each host can maintain an accurate vision of its neighbourhood, and most importantly about what kinds of documents each neighbour is interested in.

Since the neighbourhood of each host can change continuously, the information it maintains about this neighbourhood must also be updated accordingly. In practice, every time a host constructs a new announce, it forgets everything about its neighbours (see line 8 in the pseudo-code) and starts collecting "fresh" information about them. Thus, whenever it must re-construct its catalog (line 3), the profiles used to build this catalog are those of current neighbours, or more precisely those of neighbours it has heard about since the last time it updated its catalog. With this approach the catalog a host broadcast is continuously adjusted so as to fit specifically the interest profiles of its current neighbours. The cost of broadcasting an announce that contains this catalog is thus kept at a minimum: a host that maintains many documents in its local cache can avoid broadcasting blindly a large catalog on the wireless medium. Instead the catalog only pertains to documents that match its neighbours' interest profiles.

Let us consider a simple scenario for the sake of illustration. Consider the neighbour hosts $A, B$, and $C$ in Fig. 2. Assume $A$ has already received the last round of $B$'s and $C$'s periodic announces. $A$ therefore

```
<catalog>   // (built by A based on B's
            // and C's profiles)
  <descriptor>     // (excerpt from D1)
    id=''254d3g64z36cd''
    service=''filesharing''
    keywords=''mobile,ad hoc,\
              delay-tolerant,opportunistic''
    publisher=''Fred''
  </descriptor>
  <descriptor>     // (excerpt from D2)
    id=''3ab7285ef6548''
    service=''news''
    group=''comp.networking''
  </descriptor>
</catalog>
```

Fig. 10. Catalog built by *A* according to *B*'s and *C*'s profiles.

knows that *B* and *C* are (currently) its neighbours. It also knows that *B*'s interest profile is that shown in Fig. 7, and that *C*'s profile is that shown in Fig. 8. Now assume it is time for *A* to broadcast its own announce again. *A* must thus construct a catalog based on the descriptors of the documents contained in its local cache, while trying to build as small a catalog as possible. Indeed, if *A* maintains in its cache several hundreds or thousands of documents, it does not make sense to construct a large catalog containing all these documents' descriptors, if only a few of these descriptors actually match either *B*'s or *C*'s interest profile. *A* therefore parses the descriptors of the documents contained in its cache, in order to select only those descriptors that match at least one of its neighbours' profiles. Moreover, while parsing these descriptors *A* strives to select only those attributes that are distinctive selection criteria for its neighbours.

Assume *A*'s cache contains the documents whose descriptors are shown in Fig. 5. When parsing its cache looking for documents that might interest its neighbours, *A* can observe that descriptor *D2* matches the first pattern in *B*'s profile (see Fig. 7), and that in this descriptor only the attributes *service* and *group* are distinctive selection criteria for host *B*. Likewise descriptor *D1* matches the second pattern in *B*'s profile, and the attributes *service* and *keyword* are the only distinctive selection criteria for host *B*.

While considering host *C*'s profile, *A* can similarly observe that the only pattern contained in *C*'s profile (see Fig. 8) is matched by descriptor D1, although in that case the distinctive attribute is *publisher*.

Based on these observations *A* can build a catalog such as that shown in Fig. 10 (for the sake of brevity, we assume that host *A* has no other neighbours than *B* and *C*, and that its cache contains no other descriptor that matches either *B* or *C*'s interested profiles). This catalog is actually composed of excerpts of the selected documents' descriptors, since besides the *id* attribute – which must be included in each descriptor in any case – the only attributes that appear in this catalog are those that can help hosts *B* and *C* decide if they wish to receive the corresponding documents. This is consistent with our objective that the size of messages (including the periodic announce that contains a catalog) should always be kept at a minimum.

Now there are circumstances when the size of an announce can be reduced even further. As mentioned above an announce can be broadcast either in a comprehensive form or in a short form (see Fig. 9). This makes it possible for a host to avoid broadcasting a comprehensive announce, if it considers there is no point in doing so. In that case the host simply broadcasts a short-form announce, using in this announce the same hash-key as in the last comprehensive announce (note that both forms of announces include a *key* attribute, whose value is calculated based on the host's current catalog and profile, and which therefore changes only when a new comprehensive announce is constructed). By doing so it confirms its neighbours that it is still in their neighbourhood, while informing them that, from its viewpoint, the last comprehensive announce it broadcast is still valid. In practise a host can avoid building and broadcasting a new catalog when the following conditions are all verified simultaneously:

- there has been no significant change in its neighbourhood since it last broadcast a comprehensive announce (more precisely: former neighbours may have disappeared, but no new neighbour has appeared);
- the interest profiles of all known neighbours have not changed during the same interval;
- the interest profile of the announcer itself has not changed either during that interval;
- no new document has been put in the local cache during that interval.

When these conditions are all verified, a host can legitimately assume that the last comprehensive announce it has broadcast has been received and processed by its neighbours, so it can prevent from broadcasting this announce again.

### 4.1.2. Receiving a neighbour's announce

Upon receiving an announce from one of its neighbours, the receiver behaves differently depending on whether this announce is in comprehensive form or in short form.

Indeed, since all hosts in the network are assumed to move frequently (if not continuously), a host may occasionally receive a short announce from an as-yet-unknown neighbour. Moreover, because of radio interferences a host may fail to receive a comprehensive announce from one of its neighbours, and receive only a subsequent short announce from this neighbour. Our system provides for such situations. Whenever a host constructs a new comprehensive announce, this announce is also put in its local cache (line 6 in the pseudo-code), while the former announce is removed from the cache (line 2). Thus, when a host receives a short announce and realizes that it has missed the corresponding comprehensive announce, it can request that this comprehensive announce be broadcast again (line 16).

If the announce it has received is in comprehensive form, then the receiver first updates its vision of its neighbourhood based on the information (neighbour's identity and interest profile) it has just received (line 13). It then parses the catalog contained in the announce in order to identify documents whose characteristics match its own interest profile, and that are not already in its local cache (line 14). If there exists such documents, then it must actually decide which of these documents it wishes to request from the announcer. The strategy applied for managing the local cache is here of major importance, since it can influence the way the host selects these documents. Admittedly, the host may follow a greedy strategy, systematically attempting to obtain all the documents it is missing from any announcer. Yet, if the local cache is already saturated (or close to saturation), then the host must balance between the documents it is being offered, and those that are already present in its cache. Obviously it would not make sense to request many documents from a neighbour, and realize once these documents have been received that they cannot be stored in the local cache.

As mentioned in Section 2 our current objective is not to compare several cache management strategies. Instead we simply assume that each host implements a function with which it can somehow sort documents based on their sole descriptors. By applying this function to the combined set of all document descriptors it knows about (that is, those present in its local cache, plus those it is being offered by one of its

```
<request>
  from="host_id"
  to="host_id"
  docIds="254d3g64z36cd ...   3ab7285ef6548"
</request>
```

Fig. 11. Structure of a request for missing documents.

neighbours), a host can decide which of these documents it wishes to maintain in its cache and, most importantly, which of these documents it must request from its neighbour.

Once this list has been defined, the host prepares a request that simply contains the identifiers of the desired documents. This request is then sent to the announcer in a unicast control message (line 15), as shown in Fig. 11.

### 4.1.3. Processing requests

After broadcasting an announce, a host may receive requests from several of its neighbours. These requests- are processed sequentially (lines 18-22 in the pseudo-code): for each requested document, the host retrieves this document from the local cache, and broadcasts it in the network as the payload of a data message. Notice that this document is broadcast rather than being sent only to the requester in unicast mode. This is because, after broadcasting its catalog, a host may receive a series of requests for the same document (because several neighbours are interested by this single document). In such a case, all the neighbours requesting a single document from the same host can be satisfied with a single broadcast of this document. In order to avoid that consecutive requests for the same document yield a succession of re-transmissions of this document, each host maintains a history of the documents it has broadcast recently. This history is reset every time the host broadcasts a new announce. With this approach, when several neighbours ask for the same document, this document is broadcast only once in the network.

If a host receives no request after sending an announce, then it means either that it currently has no neighbour at all, or that none of its neighbours is interested in any document it can provide. Another reason might be either that the original catalog broadcast was lost, or that subsequent requests were lost, because of transient radio interferences. Such a failure is non-critical in our system, since a mobile host that misses an opportunity to exchange documents with some of its neighbours will find many other opportunities to do so in the future (and possibly with other newly found neighbours). In any case, if a host receives no request after broadcasting its catalog, then no document will be broadcast unnecessarily. This is consistent with our objective that unnecessary transmissions should be avoided, and especially transmissions of documents, which are assumed to be far larger than their descriptors.

### 4.1.4. Receiving documents

When a document is broadcast, it can be received by any neighbour of the sender. Any host that receives a document verifies if it is interested in this document by checking whether the document's descriptor matches its own interest profile (line 23). If so, then the receiver attempts to put the document in its cache, while ensuring that this is consistent with its own cache management policy.

Remember that if the host's cache is already saturated, then the host may decide that the newly received document – however interesting it might seem – is however less important than the documents present in the cache. In such a case the newly received document can be passed (if needed) to the local subscriber services, but it is not put in the local cache.

Conversely, if the host finally succeeds in storing the new document in its local cache, then it will thereafter serve as a mobile carrier for this document, thus contributing to help disseminate it further in the network.

An interesting consequence of our preferring broadcast transmissions to unicast transmissions is that it makes it possible for mobile hosts to collect documents just by *overhearing* transmissions initiated by other hosts in their neighbourhood. A host can therefore receive a document "just by chance" (that is, just because this document has been recently requested by another host). If this document is indeed an interesting one, then it can be put in the receiver's cache. The receiver has therefore obtained a document without even requesting it, and most importantly it will later refrain from requesting this document from another neighbour.

This possibility for mobile hosts to receive documents without requesting them can actually be exploited further, by allowing them to behave as *altruistic carriers* for some of these documents. The cache management policy of a host can be implemented in such a way that a host accepts to receive and store documents it is not especially interested in, provided this behaviour is obtained at low cost and does not jeopardize its prime objective, which is to collect and help disseminate interesting documents. In practice, an *altruistic* host that receives a non-interesting document can put this document in its cache only if this cache is not saturated (line 24). It will thus help disseminate this document in the network, until the cache becomes saturated and non-interesting documents must be discarded in favour to interesting ones. Note that when a host receives a non-interesting document and decides to store it in its cache for a while, there is absolutely no transmission overhead since the host simply receives "by chance" a document that has been broadcast in reply to another host's request. On the other hand, a host that carries non-interesting documents in its cache may have to send these documents to interested neighbours every now and then, which will contribute to deplete its battery. Behaving as an altruistic carrier therefore has an impact on a host's power budget. This is the reason why our system allows mobile hosts to behave as altruistic carriers, while permitting that this option be enabled or disabled on each host depending on the strategy enforced on this host.

### 4.2. Support for temporaneous message forwarding among neighbour hosts (lower layer)

As explained in Section 4.1, the upper layer of the protocol requires that a host be able to send messages (containing either an announce, a request, or a document) to its current neighbours. Temporaneous message forwarding – as opposed to delay-tolerant forwarding – is thus required in order to exploit transient connectivity between hosts that happen to reside in the same connected fragment of the network for a while. Consider for example the network shown in Fig. 1, and let us focus on the connected fragment (or island) that is marked in grey in this figure. Hosts that temporarily belong to this island can attempt to exploit the connectivity in this island in order to exchange documents through multi-hop transmissions, rather than interacting only with their direct neighbours.

Since the upper layer of our protocol requires that messages be sent either in broadcast mode or in unicast mode, the lower layer of the protocol provides support for temporaneous forwarding of unicast and broadcast messages in a connected fragment of the network.

### 4.2.1. Broadcast message forwarding

Multi-hop broadcasting in a MANET is known to be a bandwidth-consuming activity, which can occasionally lead to the so-called "broadcast storm" problem. In order to limit the overhead due to message broadcasting, the lower layer of our protocol implements a mechanism that is inspired from that used in the Optimized Link State Routing (OLSR) protocol for diffusing link-level information in the

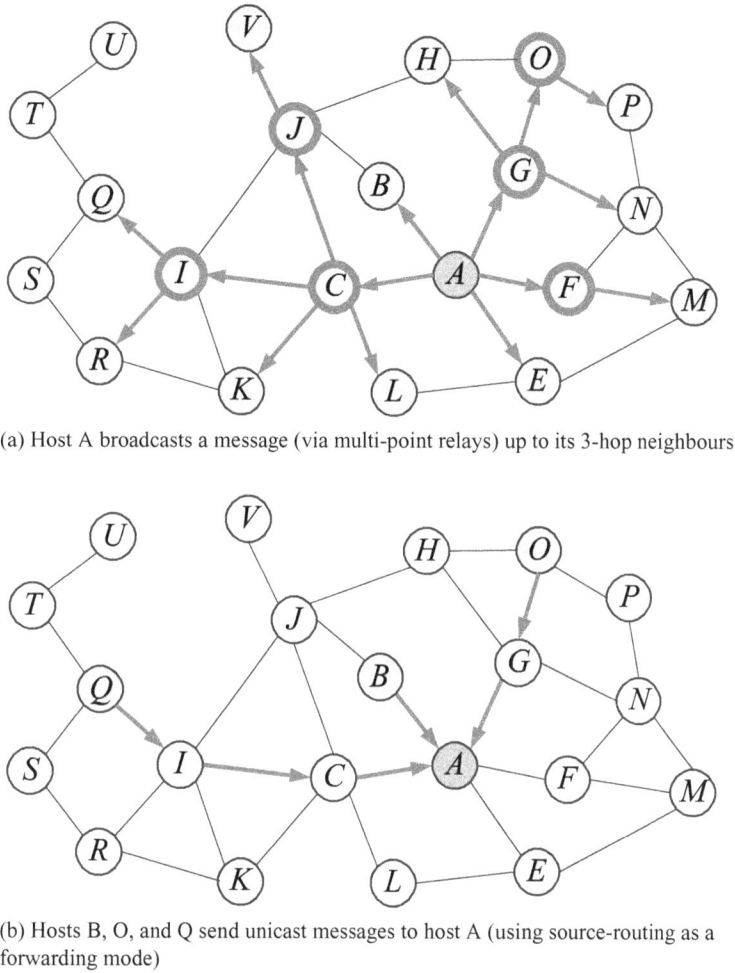

(a) Host A broadcasts a message (via multi-point relays) up to its 3-hop neighbours

(b) Hosts B, O, and Q send unicast messages to host A (using source-routing as a forwarding mode)

Fig. 12. Illustration of the two kinds of temporaneous message forwarding supported by the lower layer of our protocol.

network [6,26,18]. Basically, each node regularly selects a subset of its direct neighbours as multi-point relays (MPR), and it then relies exclusively on these MPRs for forwarding broadcast messages beyond its own radio coverage. The scope of a broadcast can be controlled by specifying how many hops a message is allowed to perform while being relayed by MPRs. Figure 12-a shows how a message can be broadcast within the greyish island shown in Fig. 1. In this example host *A* needs to broadcast a message, which could be for example an announce containing its profile and catalog. This message is allowed to propagate up to its 3-hop neighbours, but not further. Figure 13 shows the parameters inserted in each broadcast message so it can be processed and forwarded only by selected MPRs. In that case we assume that the message has already reached host *C* (one of the MPRs selected by *A*), which must now forward this message via its own MPRs *I* and *J*.

The algorithm used by each host to construct its MPR set is not detailed in this paper for the sake of brevity, and because this algorithm is very similar to that described in [26]. Basically, each host must periodically broadcast a control message in order to inform its direct (one-hop) neighbours about its presence in the network, while informing these neighbours about its own current vision of its 1-hop neighbourhood. By receiving such control messages, each host can identify its one-hop and two-hop

```
<broadcast_parameters>
  from="host_A"
  nbOfHops="2"
  mpr_set="host_I host_J"
  history="host_A host_C"
</broadcast_parameters>
```

Fig. 13. Information required in each broadcast message in order to limit its propagation scope and have it forwarded by selected MPRs only.

```
<announce>      // (Comprehensive or short form)
  ...
  1hop_neighbours="host_id1 host_id2 ..."
  ...
</announce>
```

Fig. 14. Information required for MPR selection (namely the list of 1hop neighbours) is piggy-backed in periodic announces.

neighbours, and use this information to calculate its MPR set. With the approach described in [26], specific control messages are broadcast periodically, that contain the information needed for calculating MPR sets. In our implementation, this information is piggy-backed in the announces the upper layer of the protocol must also broadcast periodically (see Fig. 14). Thus the calculation of MPR sets does not imply sending any additional message in the network: both kinds of control information (required by both layers of the protocol) are broadcast together on the wireless medium.

### 4.2.2. Unicast message forwarding

The upper layer of the protocol requires that mobile hosts be able to send requests as replies to an announce they have just received. Unicast messages must thus be forwarded towards the sender of a broadcast message. Source-routing is used as a means to perform this forwarding. Each broadcast message that propagates in the network encapsulates a history of the hosts by which it has been forwarded so far (see Fig. 13). Thus, whenever the receiver of a broadcast message decides to reply to this message, the path for sending this reply to its source is simply deduced from the path the former broadcast message has followed before reaching the receiver. Note that, in order to be effective, this approach requires that when a host decides to reply to a broadcast message, this reply is sent immediately after the broadcast message has been received. In such conditions, the path the broadcast message has followed downwards to reach the receiver is still valid in the network, so it can be followed upwards to the sender of the broadcast message.

Consider again the example shown in Fig. 12-a, and assume that hosts $B$, $Q$, and $O$ decide to reply to the message broadcast by $A$. Figure 12-b shows how their replies can propagate upwards along the path the broadcast message has just followed downwards, each reply containing a specification of the path it must follow before reaching host $A$. Figure 15 shows the parameters that must be inserted in a request sent by host $Q$ so it can be forwarded upwards to host $A$.

```
<unicast_parameters>
  from="host_Q"
  to="host_A"
  path="host_I host_C host_A"
</unicast_parameters>
```

Fig. 15. Information required in each unicast message so it can be source-routed toward its destination.

## 5. Evaluation

Our protocol for content-driven, delay-tolerant communication has been fully implemented in Java, and embedded within the DoDWAN middleware platform (as explained in Section 2). DoDWAN makes it possible to implement and experiment with different kinds of applications (such as filesharing, news distribution, messaging, etc.) in disconnected MANETs. To date it has been deployed and used extensively on up to thirty laptops with Wi-Fi capability. Yet, since it is quite difficult to run experiments with dozens or hundreds of mobile devices, DoDWAN was designed so it can also be interfaced with the MADHOC simulator [15]. Based on this combination we run a number of simulations in order to observe how the protocol can perform in different conditions, using the experience we acquired previously during real-conditions experiments to define the simulation parameters. In this section we present some of the results we obtained by performing series of 14.000 second simulation runs, with the parameters and communication scenario described below.

### 5.1. Simulation conditions

#### 5.1.1. Simulation parameters
We consider a simulation scenario in which a population of 120 users move in an environment that resembles that shown in FIg. 1. In that particular scenario, we actually consider a set of 5 buildings which are located within a 1 km × 1 km area. Each building has a rectangular shape, with edges between 100 and 150 meters long. Each user is assumed to carry a laptop equipped with an IEEE 802.11 (Wi-Fi) interface.

The mobility of users – and therefore that of the mobile hosts they are carrying – is simulated using a variant of the random waypoint model: a user can remain motionless for a while, afterwards he/she begins to walk towards a set destination, which is selected randomly in any one of the buildings in the simulation area.

In the simulation runs whose results are discussed below, we used the following mobility parameters: users are assumed to walk at speeds varying between 0.5 m/s and 2 m/s (that is, typical pedestrian speed); a stay between two consecutive moves can last between 30 seconds and 3 minutes; and the amount of intra-building mobility is set to 40% against 60% for inter-building mobility. Wi-Fi interfaces are assumed to have an omni-directional transmission range of 40 meters when used indoor, and 100 meters when used outdoor. All these parameters are consistent with observations we made while experimenting with DoDWAN in a real campus environment and with real users.

#### 5.1.2. Communication scenario.
We consider a communication scenario whereby all mobile hosts continuously produce new documents and publish these documents in the network. Each document weighs 50 kB, and each host publishes

```
<profile>
  <pattern> topic="Ti|Tj" </pattern>
</profile>
```

Fig. 16. Profile of a host interested in documents pertaining to topics $Ti$ and $Tj$.

one new document every 5 minutes. As a whole, documents are thus published in the network at an average global rate of one new document every 2.5 seconds. Topic-labelling is used as a simple means to differentiate documents: there are 16 different topics labelled $T0$ to $T15$, but each document is tagged as pertaining to only one topic.

Each mobile host is assumed to be interested in documents pertaining to only two distinct topics (hence $1/8$ of the global amount of documents published in the network). The interest profile of a host is thus defined as shown in Fig. 16. No two different hosts in the network have exactly the same interest profile.

### 5.1.3. Protocol parameters

Our protocol can be adjusted by setting two main parameters. The first parameter is the period with which a host broadcasts an announce (in either comprehensive or short form, depending on circumstances). We set this period at 15 seconds, for experience with DoDWAN in real conditions proves that this value is generally adequate in a MANET where hosts move at pedestrian speeds. Of course a shorter (resp. longer) period could be used if the hosts moved faster (resp. slower) and were expected to experience shorter (resp. longer) contacts with each other.

Another parameter is the maximum number of hops used in temporaneous message forwarding, and most notably when a host broadcasts an announce. By adjusting this parameter, we can somehow extend the "sphere of communication" of each host, controlling the scope of the announces it broadcasts periodically, and therefore the number of neighbours with which it is liable to exchange documents before moving to another part of the network.

### 5.2. Simulation results

Our first objective is to show how the scope of temporaneous message forwarding can influence the gloabl performance of our protocol. The expected result is that, when a mobile host is allowed to use multi-hop forwarding in order to interact with a large number of neighbouring hosts, documents can disseminate faster than when each host can only exchange documents with direct (one-hop) neighbours.

### 5.2.1. Speed of document dissemination

We first consider a – somewhat unrealistic – scenario where documents can propagate eternally in the network. We notably assume that the cache capacity on each host is unlimited, and that no document is given a set lifetime by its publisher. Moreover we assume, for the time being, that the option for "altruistic behaviour" (as described in Section 4) is disabled on each host, so that it is only willing to collect, carry, and forward documents that match its own interest profile.

The mobility model used during the simulation ensures that each host eventually gets close to any other host in the network. In such conditions, a document that can propagate forever in the network is guaranteed to eventually reach any interested receiver. Yet the time before this document is delivered to an interested receiver can be influenced by the protocol parameters, and notably by the scope of temporaneous message forwarding.

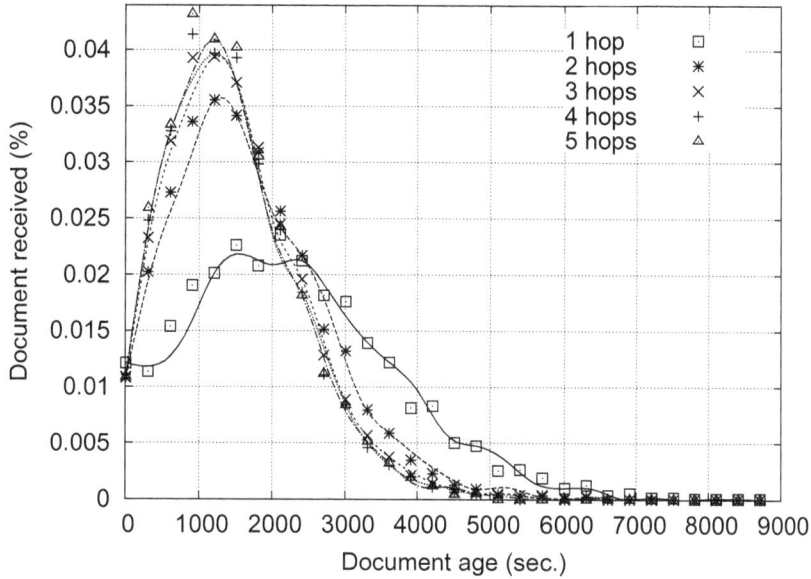

(a) Distribution of the age of documents at delivery-time

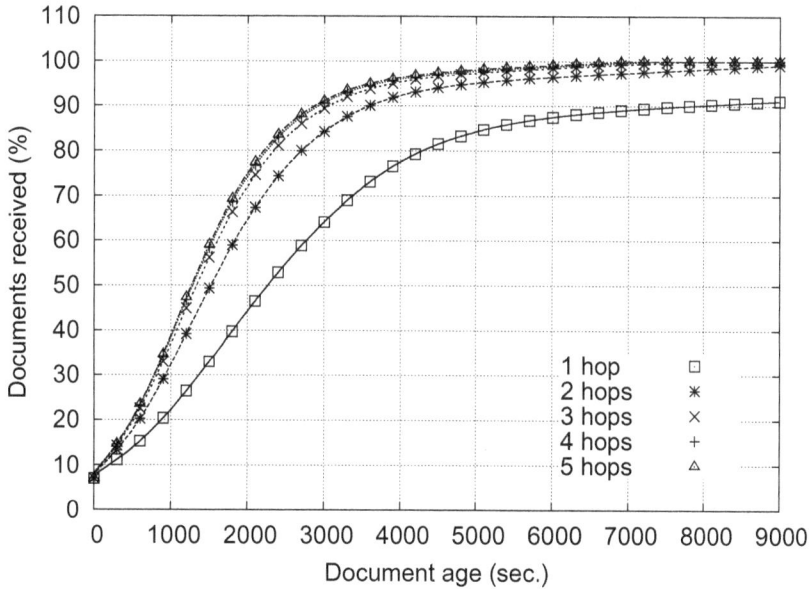

(b) Smoothed cumulative distribution of the age of documents at delivery-time

Fig. 17. Distribution and cumulative distribution of the age of documents at delivery time.

In Fig. 17 we observe how long it takes for documents to reach interested receivers. More precisely, Fig. 17-a shows the normalized distribution of the age of these documents at delivery time, and Fig. 17-b shows the corresponding cumulative distribution.

Let us first consider the case where the hosts can only use 1-hop transmissions. In such circumstances it can be observed that about 40% of the documents are delivered in less than 30 minutes. After an hour, about 75% of the documents have reached their receivers, and after two hours about 90% have been delivered.

Let us now observe how multi-hop forwarding can influence the performance of document dissemi-nation. Figure 17-a shows that, when temporaneous 2-hop forwarding is used (that is, when each host is allowed to interact with its 1-hop and 2-hop neighbours), most documents are received after about 20 minutes (against 30 minutes when only 1-hop forwarding is used). In such conditions about 98% of the documents are actually received in less than two hours, about 90% in less than an hour, and about 60% in less than 30 minutes.

A similar – though comparatively minor – improvement can be observed when multi-hop forwarding is pushed further, so that each host is allowed to extend its sphere of communication up to its 3-hop, 4-hop, and 5-hop neighbours respectively. Indeed, with the simulation parameters used during this experiment, the islands (or connected fragments of the network) that can form in the buildings have a limited extension. Their elongation varies between 0 (isolated hosts) and 7 hops, with an average value of 4.2 hops. This explains why extending the sphere of communication of each host beyond a couple of hops does not bring much improvement. Another reason is that the propagation of documents between different buildings (or between non-connected parts of a building) depends primarily on how fast document carriers – that is, pedestrians in the scenario considered – actually move in the simulation area.

In any case, this first experiment confirms that by extending the sphere of communication of each mobile host our protocol allows documents to disseminate better and faster in each island, thus increasing the number of hosts that can then serve as carriers between non-connected parts of the network.

### 5.2.2. Cache capacity

In the simulation runs whose results were discussed above, we assumed that documents could propagate forever in the network. As mentioned above this is not very realistic, since most resources in a MANET are usually severely constrained. For example the cache where mobile hosts can store documents is of limited capacity. An adequate policy must thus be devised – and then enforced on each host – in order to deal with saturation conditions.

Figure 18 shows how the capacity of each host's cache can influence the performance of document dissemination. To obtain these results we run a series of simulations, considering cache capacities ranging between 50 and 200 documents. During each simulation the cache policy enforced was such that, when a cache reached saturation, the oldest document in this cache was discarded in order to make room for a new document. In the figure we plot the satisfaction ratio (that is, the percentage of documents that are eventually delivered to interested receivers) against the capacity of the cache. First, Fig. 18 confirms the natural expectation that a host with a larger cache is liable to carry documents further and longer in the network.

More interesting is the influence of temporaneous multi-hop forwarding on the performance of doc-ument dissemination. In Fig. 18 it can be observed that the satisfaction ratio of document delivery increases significantly when the scope of message forwarding is extended to a couple of hops around each host. Consider for example the case where each host can only maintain 100 documents in its cache. In such conditions, the documents sent in the network are received (on average) by only 78% of the interested receivers if each host is only allowed to interact with direct neighbours. Yet this figure is increased by 10% when the scope of temporaneous forwarding is extended to 2-hop neighbours, and again by 2% when it is extended to 3-hop neighbours.

### 5.2.3. Document lifetime

Another way to prevent documents from remaining eternally in the hosts' caches is to give each document a set lifetime, so that whenever a document gets obsolete it is automatically removed from any

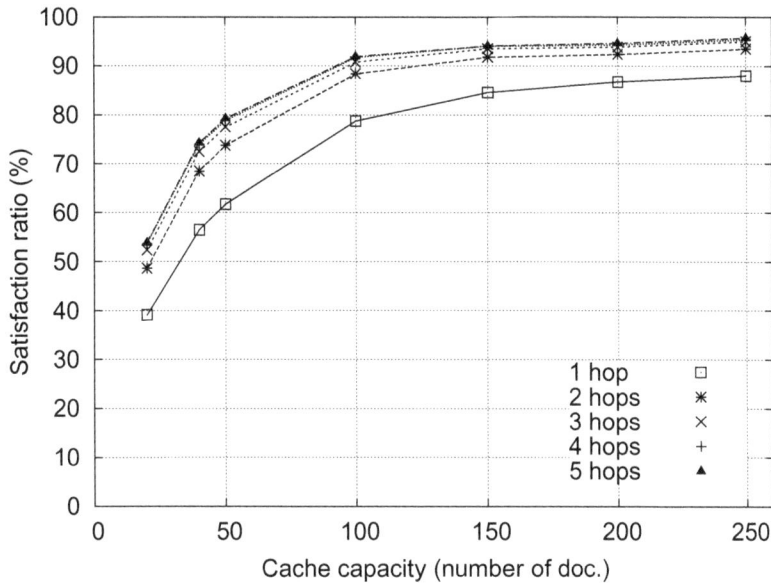

Fig. 18. Satisfaction ratio (of document delivery) vs. cache capacity.

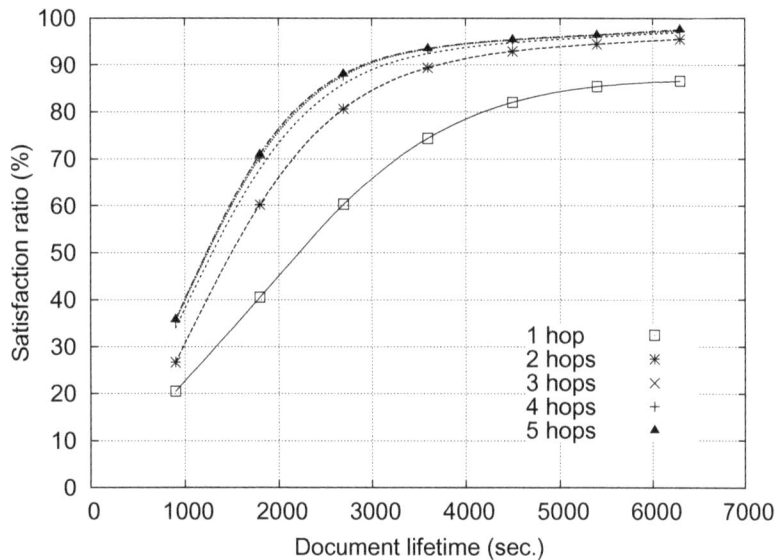

Fig. 19. Comparison of the MPR-based and flooding-based versions of the protocol.

cache it might have been stored in. This method can be used either as a substitute or as a complement to the method that limits the capacity of each cache.

Figure 19 shows how different values of document lifetime influence the performance of document dissemination. These results were obtained with unbounded cache capacity, so that the two types of constraints do not interfere during the simulation. In the figure we plot the satisfaction ratio (percentage of documents that are delivered to interested receivers) against the set lifetime of documents. Not surprisingly, the satisfaction ratio increases as documents are given a longer lifetime. Yet it can again be observed that temporaneous multi-hop forwarding gives significant improvement in document dissemination. For example, when documents are given a 30-minute lifetime, they are eventually received (on average) by only 40% of the interested receivers if each host is only allowed to interact with direct neighbours. Yet this figure is increased by 20% when the scope of temporaneous forwarding is extended to 2-hop neighbours, and again by 7% when it is extended to 3-hop neighbours.

### 5.2.4. Communication overhead

The above results confirm that by resorting to temporaneous multi-hop forwarding, the dissemination of documents in the network can be made faster, and thus more efficient. They also show that even a slight extension of the sphere of communication of each host (by only two or three hops in the scenario considered) can bring a significant improvement over a situation where a host can only interact with direct neighbours.

The drawback of multi-hop forwarding is that it yields an important overhead in terms of the resources it mobilises on each host. Indeed, whenever a host forwards a message, this transmission drains the battery of this host, while occupying the shared wireless medium around this host.

While designing our protocol we decided to rely on MPR-based forwarding for broadcasting messages around each host. Obviously it would have been a lot easier for us to use plain flooding for broadcasting these messages. Since above-mentioned results show that messages need only be forwarded on a limited scope (typically, two or three hops) it is worth wondering whether MPR-based forwarding brings any benefit over plain flooding in such conditions.

In order to evaluate the difference between our approach relying on multi-point relays and an alternate one relying on plain flooding, we implemented a variant of our protocol that uses plain flooding as a means to broadcast messages around each sender. The results are presented in Fig. 19. They were obtained when running our communication scenario during four hours (in simulation time), with unlimited cache capacity and 1-hour document lifetime.

It can be observed (Fig. 19-a) that the MPR-based and flooding-based versions of the protocol do not give exactly the same satisfaction ratio. This is because the MPR-based version is slightly slower at disseminating documents in the network. Indeed, with this version a host whose neighbourhood changes needs to wait a while (precisely, two consecutive announce cycles) because it can effectively interact with its new neighbours. In contrast, with the flooding-based version of the protocol a host whose neighbourhood changes can immediately reach its new neighbours.

The satisfaction ratio observed with the MPR-based version of the protocol is therefore slightly lower than with the flooding-based protocol. Yet this difference remains under 3%, while the cost of using one or the other way of broadcasting messages is very different. Figure 19-b shows how the cost of transmissions compares with both versions of the protocol. Obviously our decision to rely on multi-hop relays for forwarding broadcast messages is fully justified, as the global number of messages sent when using multi-hop relays is far below that observed when flooding messages in the network.

### 5.2.5. Adaptive catalog

In Section 4 we have claimed that our protocol has been designed so as to consume as little resources as possible. We have notably described how the catalog each host inserts in its periodic announces is constructed so as to match exactly the interest profiles of its current neighbours. Figure 20 shows how the size of the catalog broadcast by a particular host evolves over time, depending on whether this host actually has neighbours, depending on these neighbours' interest profiles, and depending of course on the documents it already maintains in its cache. The results presented in Fig. 20 were obtained during

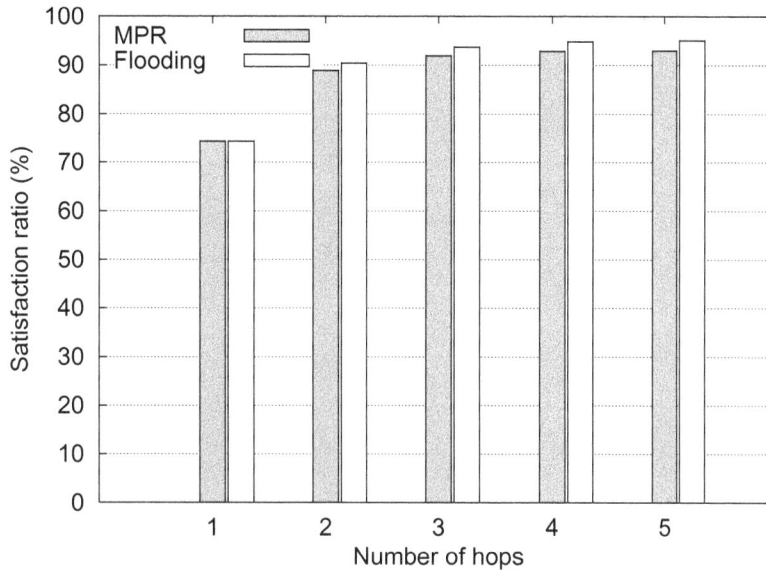

(a) Satisfaction ratio.

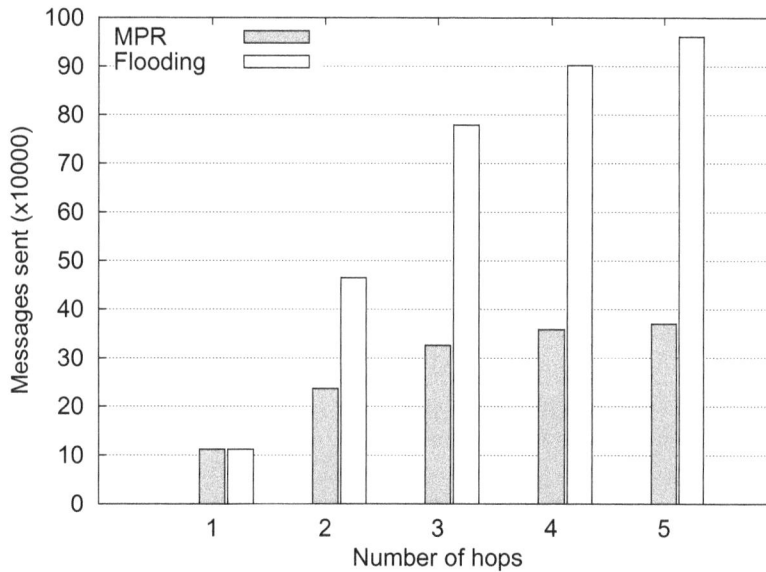

(b) Transmission cost.

Fig. 20. Evolution of catalog size and cache occupancy over time on a single mobile host.

a simulation where the capacity of the cache on the considered host was set to 100 documents. Each document was given a 75 minute lifetime, and the scope of multi-hop forwarding was set to 2 hops.

In the figure we can first observe that, during the interval considered, the cache is almost continuously full. Cached documents are discarded as soon as they become obsolete, but new documents obtained from neighbour hosts fill in the gap soon afterwards. Yet the number of descriptors inserted in the catalog the host constructs periodically (at most every 15 seconds) is often smaller than the number of documents it maintains in its cache, and sometimes falls down to 0 (empty catalog). This is because the host builds its catalog by selecting only documents that can interest its neighbours. Sometimes the host has no

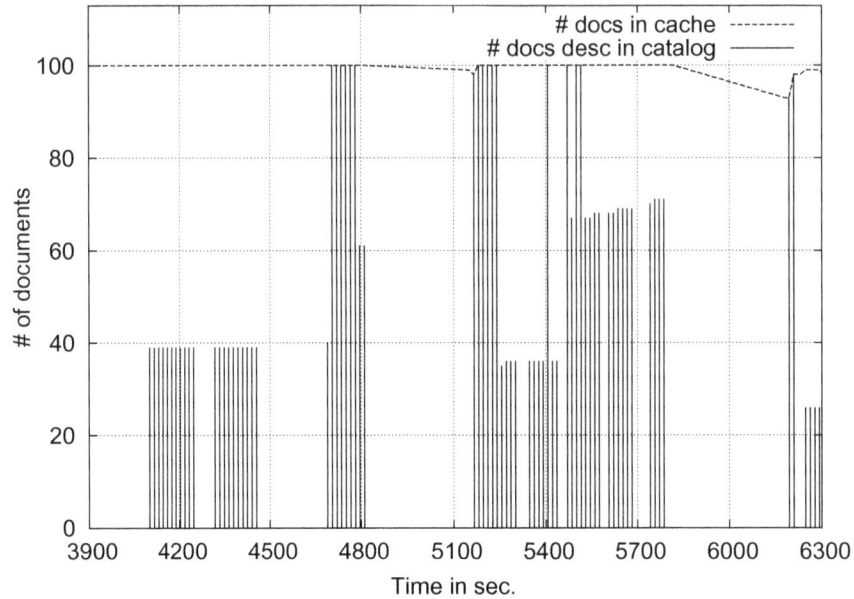

Fig. 21. Evolution of catalog size and cache occupancy over time on a single mobile host.

neighbour at all, sometimes its neighbours have interest profiles that do not intersect its own, so that it cannot propose them any document they might be interested in. Sometimes its neighbours are interested in only a small subset of the documents it maintains in its cache, so the catalog only concerns this subset.

These results confirm that by adapting continuously its catalog based on its neighbours' profiles, a host can contribute to reduce the weight of its periodic announces and, more generally, the amount of work expected from any neighbour that must receive and analyse these announces. Of course, our efforts for reducing the size of each host's catalog proves even more effective when each host maintains a large number of documents in its cache, while its neighbours present highly selective interest profiles. In some simulation scenarios (not detailed here) we have considered hosts capable of maintaining up to 10.000 documents in their cache, whereas the selectivity of their neighbours' profiles was such that only a very small fraction (actually less than 1%) of these documents had to be proposed in each catalog. The fine-tuning of each catalog proves very profitable in such circumstances.

*5.2.6. Altruistic behaviour*

In Section 4 we have explained how each host can be configured so as to behave as an altruistic carrier for documents it is not especially interested in, without compromising its chance of collecting documents that match its own interest profile. Yet this possibility has not been used in the simulation runs whose results have been presented so far. Let us now observe how the global performance of document dissemination is affected when hosts are allowed to behave altruistically.

In Fig. 22 we observe how the satisfaction ratio of document delivery evolves over time, depending on whether the mobile hosts adopt either an altruistic or a selfish behaviour. A host is said to behave selfishly when it only accepts to store, carry, and forward only documents it is itself interested in. This corresponds to the behaviour we have actually considered in al lthe results presented so far. A host is said to behave altruistically when it accepts to receive and store in its cache documents it overhears on the wireless medium, even though these documents present no interest to him. The results presented in Fig. 22 were obtained during a simulation where the capacity of each host's cache was set to 300 documents, which in

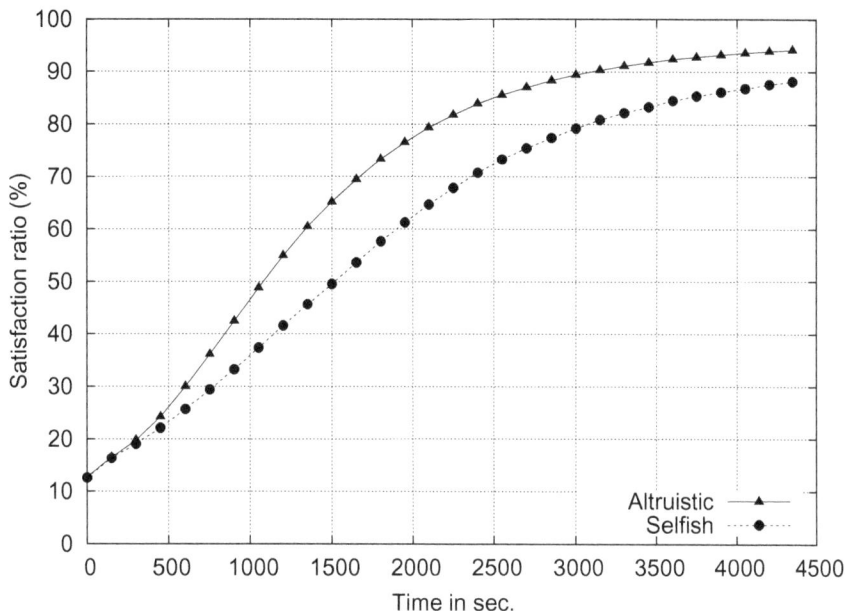

Fig. 22. Satisfaction ratio (of document delivery) over time when mobile hosts behave either altruistically of selfishly.

that particular case is slightly larger than the capacity required for carrying only interesting documents, and therefore allows that the remaining space be used for carrying non-interesting ones. Each document was given a 75 minute lifetime, and the scope of multi-hop forwarding was set to 2 hops.

In this figure we can observe that when the hosts are allowed to behave as altruistic carriers, documents can indeed disseminate faster – and therefore more efficiently – in the network. For example, with selfish hosts the documents can be received on average by 57% of all interested subscribers in less than 30 minutes, whereas with altruistic hosts this figure is about 72% (hence a 15% improvement).

Of course, this observation confirms again the natural expectation that documents disseminate better when they can be transported by a larger number of mobile carriers. Yet it is worth recalling that, with our approach, this improvement comes at very little cost, since each host basically collects non-interesting documents by overhearing their transmission on the wireless medium, and since an altruistic host never removes an interesting document from its cache in order to make room for a non-interesting one.

## 6. Related work and discussion

The concept of content-based networking has originally been introduced in [5]. Since then it has been refined in a number of papers such as [3,4]. In [3] the authors notably propose several levels of predicate languages that can be used to filter messages based on their content, including languages that apply regular expressions either to attribute names or to flat messages. In our system, document differentiation currently relies on a rather simplistic model: differentiation is performed by comparing document descriptors only (rather than the whole content of these documents), and a subscriber's selection predicate is defined as a conjunction of regular expressions that only apply to attribute values. Improving the expressiveness of this model is one of our objectives in the near future.

Many papers have been published in the last few years that address the problem of supporting communication in disconnected MANETs [24,32]. Some of these papers actually assume that mobility patterns are known in advance or can be controlled as needed (e.g. [19,33]), while others make no such assumption and propose to rely on redundancy in order to improve the reliability of delay-tolerant transmission. In the latter category, it is usually proposed to rely on more or less controlled forms of epidemic or probabilistic propagation schemes [11,23,27–29]. Some papers specifically consider communication between user-carried devices, and propose to drive message forwarding in the network by predicting how users move or meet, or by identifying what communities each user belongs to [17]. Indeed, in these papers the basic assumption is that users tend to exhibit regular mobility and/or social interaction patterns, which can be identified (more or less automatically) and then used to select the best carriers for messages addressed to a particular user. For example [20] defines a probabilistic approach whereby the probability to deliver a message to its destination is calculated based on a delivery predictability metric that is derived from the history of node encounters. Similarly, [22] attempts to predict node contacts, using a model of prediction over time series that allows to forecast co-location probability. [2] proposes a context-based approach, whereby each host must maintain a history of context information pertaining to each host (or user) it has encountered in the past. Whenever a message is sent in the network the sender must provide meta-information about the destination (such as the recipient's residence or work address), so this information can be matched against that available in each potential carrier's history in order to calculate delivery probabilities [16]. Describes a protocol for social-based forwarding, whereby user communities are identified automatically (using an approach similar to that described in [17]), and users that belong to the same community as a message recipient are selected as best carriers for that message.

In most of the above-mentioned papers the objective is to reach a set destination, specified by the sender. In contrast in content-based communication the sender does not necessarily know who the recipients of its message are, or even if they exist at all. Several papers about content-based communication have already been published, but the algorithms and protocols they define can only be used in stable, wired networks, or in fully connected MANETs [9,13,21,25]. These papers usually propose to construct and maintain content-based *routing structures* in order to forward messages efficiently between publishers and subscribers. A notable exception with that respect is the protocol defined in [1]. Like ours this protocol does not attempt to build any structure to support routing decisions. Instead it too relies on broadcast transmissions, while deferring to hosts that receive a message the decision to forward this message to potential subscribers, based on an estimation of their distance to these subscribers. Yet this protocol requires that temporaneous end-to-end paths exist between senders and receivers. It could not run satisfactorily in a disconnected MANET.

Content-based dissemination in disconnected MANETs is addressed specifically in [8], which describes an approach whereby a content-driven multi-hop routing structure (limited to a given horizon) is built around each host. A utility-based function is used in order to select the best forwarders for each kind of message, and mobile carriers help disseminate messages between non-connected parts of the network. Our protocol relies on a slightly different approach. Instead of attempting to construct and maintain a routing structure, it relies on periodic broadcast transmissions (also limited to a given "horizon" from the sender), whereby each host periodically informs its neighbours about the documents it is carrying and that match their interest profiles. Upon receiving such a catalog a host can request the transmission of a document it is actually missing. Thus no document is sent in the network unless it has been requested explicitly by a client host.

[10] defines the Autonomous Gossipping (A/G) algorithm, that allows neighbour hosts to opportunistically exchange documents they are missing, based on their respective advertised profiles. In the A/G algorithm, information dissemination is actually depicted as an epidemic process: each host is

considered as being more or less vulnerable to being "infected" by one or another kind of data item. One difference between our protocol and the A/G algorithm is that the latter only relies only on direct interactions between one-hop neighbours, whereas ours supports interaction in connected fragments of the network through multi-hop transmissions. Simulations show that this possibility for a host to reach n-hop neighbours makes the dissemination of information more effective when islands actually appear in the network, as it helps compensate for the selectivity of each host's interest profile. Moreover, to the best of our knowledge the A/G algorithm was never actually implemented (except as a simulator), whereas our protocol has been fully implemented in a middleware platform, so it can now run either in real conditions, or coupled to a simulator.

[31] proposes to use a clustering algorithm to create a Publish/Subscribe overlay, in which centrality nodes somehow behave as brokers between message publishers and subscribers. [7] describes a protocol for Publish/Subscribe that is derived from the protocol for unicast routing presented in [22], but that can account for the interests of users. This protocol exploits predictions based on metrics of social interactions to identify the best message carriers. In fact, the basic assumption is that users with common interests tend to meet with each other more often than with other users. The routing algorithm exploits this property by selecting as carriers for messages hosts which have often been co-located with interested subscribers in the past.

Admittedly there are circumstances when people with similar interests tend to meet regularly. This is for example the case when these people are co-workers (or fellow students), or when they are members of the same family, the same sports club, or the same game club. Yet there are also cases when the fact that people share similar interests does not imply that they are members of the same closely-knit – or even loosely-knit – community. For example people with a keen interest in football or rugby do not necessarily meet very often. Indeed some of them meet frequently in stadiums, but many others simply watch matches on TV. People who wish to keep informed about weather forecasts or about weekly TV programs do not necessarily meet very often either. Similitude between their interest profiles simply means that they can share information occasionally, but this similitude is hardly correlated with their mobility or co-location pattern. In such conditions it is very unlikely that any history-based approach (that basically attempts to predict future movements and/or contacts between users based on an observation of past mobility and/or contacts) can prove very efficient.

Another problem with history-based prediction techniques is that they do not scale up very well, because of the overhead implied by history maintenance. Imagine for example that every citizen in a medium-size city – say 20.000 inhabitants – carries a digital device capable of short-range, ad hoc communication (such as a PDA or smart-phone with a built-in Wi-Fi interface). To the best of our knowledge, it is still unclear whether history-based algorithms, in which each host must continuously collect and maintain data about any other host it meets while roaming the network, can scale up to such a large network.

It is our conviction that although history-based prediction techniques can prove very efficient in small networks, there is also a need for systems that can run in larger networks. We claim that the system we designed can indeed run in a large disconnected MANET, as it does not attempt to build any history of a mobile host's encounters with other hosts. Indeed, in our system each host only maintains very little information about its neighbours, and forgets everything about them as soon as its neighbourhood changes. The main limitations of the system are therefore the number of neighbours a host can have at any time (although this constraint mostly depends on the characteristics of the underlying wireless technology), and the number and size of the documents it can exchange with neighbours while they are co-located.

In order to prove our claim that our system can indeed perform satisfactorily in a large network, it would be most interesting to run simulations with a large number of mobile hosts. Unfortunately, the simulator we used for evaluating our system is not distributed and must therefore run on a single workstation. The CPU speed and memory available on this workstation are thus the limiting factors during simulations. With this simulator we could actually run simulations with up to 10.000 hosts, but in that case each host could only maintain a very small cache, so the number of documents disseminated during these simulations was not very impressive. We could also run simulations with more than 100.000 documents disseminating in the network, but then it is the number of hosts that was limited. Raising this constraint, so we can simulate realistic scenarios in large networks, is an important item in our agenda.

## 7. Conclusion

In this paper we have presented a new system for content-based communication in disconnected MANETs. Unlike other protocols that rely on costly methods for constructing and maintaining content-driven routing structures, ours does not attempt to build any such structure. Instead it exploits transient contacts between mobile hosts that get close enough to one another, allowing these hosts to exchange documents according to their respective interest profiles. Communication between non-connected fragments of the network is performed thanks to mobile hosts, each host serving as a carrier for documents it maintains in a local cache. In our system a host is primarily expected to collect and carry information it is itself interested in, but it can also behave as an altruistic carrier for non-interesting documents, as long as this behaviour does not compromise its chance of collecting interesting ones. Simulation shows that our protocol is effective at propagating documents between senders and interested receivers. Its use of temporaneous multi-hop forwarding helps disseminate documents in connected fragments of the network, which in turn has a positive influence on this dissemination in the whole, disconnected network. By adjusting the extension of multi-hop forwarding around each host, the resulting transmission overhead can be balanced against the benefit observed in document dissemination. With the current version of the protocol the number of hops used when broadcasting messages is set as a constant parameter. In the future we plan to investigate methods allowing each host to adjust this value dynamically, accounting for its current situation in the network (e.g. number and density of neighbours, interest profiles of these neighbours, history of recent document exchanges in the neighbourhood, etc.). We also consider improving the way document differentiation is achieved by allowing more elaborate forms of attribute filtering.

## Acknowledgements

This work is supported by the French *Agence Nationale de la Recherche* under contract ANR-05-SSIA-0002-01. It is also supported by the French Armament Procurement Agency (DGA) by means of a Ph.D. grant. The authors would like to thank the reviewers of the original version of this paper (submitted for the AINA'08 conference) for their helpful comments and suggestions.

## Appendix: pseudo-code of the content-based dissemination protocol

### Variables

$self$: node's own identifier.

$profile$: node's own interest profile.

$C$: node's document cache.

$announce$: node's last comprehensive announce.

$neigh < key, prof > [i]$: contains a tuple composed of the last hash-key, profile, and catalog received from node $i$.

$S$: set of identifiers of documents that have been broadcast recently.

$altruistic$: boolean flag. True if the node's altruistic behaviour is enabled.

### Messages

ANNOUNCE$< n, key, prof, cat >$: comprehensive form of the announce a host can send periodically. It contains the host's id, a hash-key, and the node's profile and catalog.

ANNOUNCE$< n, key >$: short form of the announce a host can send periodically. It only contains the node's id and a hash-key (whose value is the same as that of the last comprehensive announce sent by this host).

REQUEST$< docIds >$: message requesting the broadcast of the documents whose ids are specified in list $docIds$.

DOCUMENT$< desc, data >$: message containing a document (descriptor and data).

### Functions

**broadcast**(m): broadcast message $m$.

**send**(m, n): send message $m$ to destination $n$.

**hashKey**(c, p): compute hash-key based on a host's catalog $c$ and profile $p$.

**createCatalog**(neigh, C): create a catalog by selecting in cache $C$ documents that match the current neighbours' profiles.

**identifyMissingDocs**(c, p, C): process catalog $c$ based on profile $p$, and return a list of identifiers of interesting documents mentioned in $c$ that are not already in cache $C$.

**conditionsHaveChanged**: return true if something has changed that justifies updating the host's announce (i.e. at least one new neighbour has been discovered, a neighbour's profile has changed, a new document has been put in the local cache, or the node's profile has changed).

— Periodic announce —

**send**(ANNOUNCE)

1  **if conditionsHaveChanged**() **then**
    *// Prepare a comprehensive announce*
    *// (and store a copy in cache)*
2      $\mathscr{C} \leftarrow \mathscr{C} \setminus \{announce\}$
3      cat ← **createCatalog**(neigh, $\mathscr{C}$)
4      *key* ← **hashKey**(*cat, profile*)
5      *announce* ← ANNOUNCE<self, key, profile, cat>
6      $\mathscr{C} \leftarrow \mathscr{C} \cup \{announce\}$
7      *msg* ← *announce*
8      *neigh* ← $\emptyset$
9  **else**
    *// Prepare a short announce*
10     *msg* ← ANNOUNCE<self, key>
11  $\mathscr{S} \leftarrow \emptyset$
12  **broadcast**(*msg*)


— Invoked on receipt of a comprehensive announce —

**receive**(a: ANNOUNCE<n, key, prof, cat> )

13  neigh[n]=<key,prof>
14  docIds ← **identifyMissingDocs**(cat, profile, $\mathscr{C}$)
15  **send**(REQUEST<docIds>, n)


— Invoked on receipt of a short announce —

**receive**(a: ANNOUNCE<n, key>)

16  **if** *neigh[n].key != key* **then**
    *// Failed to receive the comprehensive version*
    *// of this announce. Requesting one.*
17     **send**(REQUEST<key>, n)


— Invoked on receipt of a request —

**receive**(r: REQUEST<docIds>)

18  **forall** *id* $\in$ *docIds* **do**
19     **if** *id* $\notin$ $\mathscr{S}$ **then**
        *// A document is sent only once during a period*
20         DOCUMENT *d* ← $\mathscr{C}$.get(id)
21         $\mathscr{S} \leftarrow \mathscr{S} \cup \{id\}$
22         **broadcast**(*d*)


— Invoked on receipt of a document —

**receive**(d: DOCUMENT<desc,data>)

23  **if** *((desc.matches(profile)* $\wedge$ *d* $\notin$ $\mathscr{C}$)
24  **or** *(altruistic* $\wedge$ $\mathscr{C}$.notFull()))* **then**
25     $\mathscr{C} \leftarrow \mathscr{C} \cup \{d\}$


**Algorithm 1**. protocol functions.

# References

[1] R. Baldoni, R. Beraldi, M. Migliavacca, L. Querzoni, G. Cugola and L. Migliavacca, Content-Based Routing in Highly Dynamic Mobile Ad Hoc Networks, *Journal of Pervasive Computing and Communication* **1**(4) (Dec 2005), 277–288,

[2] C. Boldrini, M. Conti, J. Jacopini and A. Passarella, HiBOp: a History Based Routing Protocol for Opportunistic Networks. In *Proceedings of the IEEE International Symposium on a World of Wireless*, Mobile and Multimedia Networks, IEEE, June 2007, pp. 1–12.

[3] A. Carzaniga and C.P. Hall, Content-Based Communication: a Research Agenda. In *Proceedings of the 6th International Workshop on Software Engineering and Middleware*, ACM, Nov 2006, pp. 2–8.

[4] A. Carzaniga, M.J. Rutherford and A.L. Wolf, A Routing Scheme for Content-Based Networking. In *Proceedings of IEEE INFOCOM 2004*, IEEE, Mar 2004, pp. 918–928.

[5] A. Carzaniga and A.L. Wolf, Content-based Networking: a New Communication Infrastructure. In *Proceedings of the NSF Workshop on an Infrastructure for Mobile and Wireless Systems*, number 2538 in LNCS, Springer, Oct 2001, pp. 59–68.

[6] T. Clausen and P. Jacquet, Optimized Link-State Routing Protocol (OLSR). IETF, RFC 3626, Oct 2003.

[7] P. Costa, C. Mascolo, M. Musolesi and G.P. Picco, Socially-Aware Routing for Publish-Subscribe in Delay-Tolerant Mobile Ad Hoc Networks, *IEEE Journal On Selected Areas In Communications (JSAC)* **26**(5) (June 2008), 748–760.

[8] P. Costa, M. Musolesi, C. Mascolo and G.P. Picco, Adaptive Content-based Routing for Delaytolerant Mobile Ad Hoc Networks. Technical report, UCL, Aug 2006.

[9] P. Costa and G.P. Picco, Semi-Probabilistic Content-Based Publish-Subscribe. In *25th International Conference on Distributed Computing Systems (ICDCS 2005)*, IEEE, June 2005, pp. 575–585.

[10] A. Datta, S. Quarteroni and K. Aberer, Autonomous Gossiping: a Self-Organizing Epidemic Algorithm for Selective Information Dissemination in Mobile Ad-Hoc Networks. In *Semantics for Grid Databases, First International IFIP Conference (ICSNW 2004)*, number 3226 in LNCS, Springer, June 2004, pp. 126–143.

[11] P. Eugster, R. Guerraoui, A.-M. Kermarrec and L. Massoulié, From Epidemics to Distributed Computing, *IEEE Computer* **37**(5) (May 2004), 60–67.

[12] K. Fall, A Delay-Tolerant Network Architecture for Challenged Internets. In *Proceedings of the 2003 conference on Applications*, technologies, architectures, and protocols for computer communications, ACM, Aug 2003, pp. 27–34.

[13] U. Farooq, S. Majumdar and E. Parsons, High Performance Publish/Subscribe Middleware for Mobile Wireless Networks, *International Journal of Mobile Information Systems* **3**(2) (2007), 107–132.

[14] K.A. Harras, K.C. Almeroth and E.M. Belding-Royer, Delay Tolerant Mobile Networks (DTMNs): Controlled Flooding in Sparse Mobile Networks. In *IFIP Networking Conference, Waterloo, Ontario, CANADA*, volume 3462 of LNCS, Springer, May 2005, pp. 1180–1192.

[15] L. Hogie, P. Bouvry and F. Guinand, The MADHOC simulator. http://www-lih.univlehavre.fr/ hogie/madhoc.

[16] P. Hui, J. Crowcroft and E. Yoneki, BUBBLE Rap: Social Based Forwarding in Delay Tolerant Networks. In *Proceedings of the 9th ACM International Symposium on Mobile Ad Hoc Networking and Computing*, ACM, May 2008, pp. 241–250.

[17] P. Hui, E. Yoneki, S.-Y. Chan and J. Crowcroft, Distributed Community Detection in Delay Tolerant Networks. In *Proceedings of the 2nd ACM/IEEE international workshop on Mobility in the evolving internet architecture*, ACM, Aug 2007, pp. 1–8.

[18] M. Ikeda, L. Barolli, G. De Marco, T. Yang, A. Durresi and F. Xhafa, Tools for Performance Assessment of OLSR Protocol, *International Journal of Mobile Information Systems* **5**(2) (2009), 165–176.

[19] Q. Li and D. Rus, Sending Messages to Mobile Users in Disconnected Ad-hocWireless Networks. In *Proceedings of the Sixth Annual International Conference on Mobile Computing and Networking*, Aug 2000, pp. 44–55.

[20] A. Lindgren, A. Doria and O. Schelen, Probabilistic Routing in Intermittently Connected Networks. In *Proceedings of the First International Workshop on Service Assurance with Partial and Intermittent Resources (SAPIR 2004)*, volume 3126 of LNCS, Springer, Aug 2004, pp. 239–254.

[21] R. Meier and V. Cahill, STEAM: Event-Based Middleware for Wireless Ad Hoc Network. In *International Conference on Distributed Computing Systems, Workshops (ICDCSW '02)*, IEEE, July 2002, pp. 639–644.

[22] M. Musolesi, S. Hailes and C. Mascolo, Adaptive Routing for Intermittently ConnectedMobile Ad Hoc Networks. In *Proceedings of the IEEE 6th International Symposium on a World of Wireless, Mobile, and Multimedia Networks*, IEEE, June 2005, pp. 183–189.

[23] M. Musolesi, C. Mascolo and S. Hailes, Emma: Epidemic messaging middleware for ad hoc networks, *Personal Ubiquitous Computing* **10**(1) (Aug 2005), 28–36.

[24] L. Pelusi, A. Passarella and M. Conti, Opportunistic Networking: Data Forwarding in Disconnected Mobile Ad Hoc Networks, *IEEE Communications Magazine* **44**(11) (Nov 2006), 134–141.

[25] M. Petrovic, V. Muthusamy and H.-A. Jacobsen, Content-Based Routing in Mobile Ad Hoc Networks. In *Proceedings of the 2nd Annual International Conference on Mobile and Ubiquitous Systems: Networking and Services (MobiQuitous'05)*, IEEE, July 2005, pp. 45–55.

[26]   A. Qayyum, L. Viennot and A. Laouiti, Multipoint Relaying for Flooding Broadcast Messages in Mobile Wireless
        Networks. In *Proccedings of the 35th Annual Hawaii International Conference on System Sciences (HICSS'02)*, IEEE,
        Jan 2002, pp. 3866–3875.

[27]   Y. Sasson, D. Cavin and A. Schiper, *Probabilistic Broadcast for Flooding in Mobile Ad Hoc Networks*, Technical Report
        IC/2002/54, Swiss Federal Institute of Technology (EPFL), 2002.

[28]   T. Spyropoulos, K. Psounis and C.S. Raghavendra, Spray and Wait: an Efficient Routing Scheme for Intermittently
        Connected Mobile Networks. In *2005 ACM SIGCOMM workshop on Delaytolerant networking (WDTN'05)*, ACM, Aug
        2005, pp. 252–259.

[29]   A. Vahdat and D. Becker, *Epidemic Routing for Partially Connected Ad Hoc Networks*, Technical report, Duke University,
        Apr 2000.

[30]   E. Vollset, K. Birman and R. van Renesse, Chickweed: Group Communication for Embedded Devices in Opportunistic
        Networking Environments. In *3rd InternationalWorkshop on Dependable Embedded Systems, in conjunction with 25th
        Symposium on Reliable Distributed Systems (WDES 2006)*, Oct 2006.

[31]   E. Yoneki, P. Hui, S.-Y. Chan and J. Crowcroft, A Socio-Aware Overlay for Publish/Subscribe Communication in Delay
        Tolerant Networks. In *Proceedings of the 10th ACM/IEEE International Symposium onModeling, Analysis and Simulation
        ofWireless andMobile Systems (MSWiM)*, ACM, Oct 2007, pp. 225–234.

[32]   Z. Zhang, Routing in Intermittently Connected Mobile Ad Hoc Networks and Delay Tolerant Networks: Overview and
        Challenges, *IEEE Communications Surveys and Tutorials* **8**(1) (Jan 2006), 24–37.

[33]   W. Zhao, M. Ammar and E. Zegura, A Message Ferrying Approach for Data Delivery in Sparse Mobile Ad Hoc Networks.
        In *Proceedings of ACM Mobihoc 2004*, May 2004, pp. 187–198.

**Julien Haillot** obtained his Master's degree at the Université de Bretagne Sud (France) in 2006. He is presently preparing his
Ph.D. degree in Computer Science under the supervision of Dr. Frédéric Guidec in the VALORIA laboratory at Université de
Bretagne Sud, with a research grant delivered by the French Armament Procurement Agency. His research interests pertain
to mobile, opportunistic, and delay/disruption-tolerant ad hoc networking, with a special focus on content-based information
dissemination in military tactical networks.

**Frédéric Guidec** received his Ph.D. in Computer Science from the Université de Rennes I (France) in 1995. He then joined the
Swiss Federal Institute of Technology in Lausanne (Switzerland), where he took part in several projects involving distributed
and parallel computing for scientific numerical computation. He is now an associate professor at the Université de Bretagne Sud
(France). His present major research line concerns mobile ad hoc networking in challenged environments. He is the author of
over 50 papers dealing with parallel, distributed, and bio-inspired computing, radio network planning, context-aware software,
and opportunistic or delay/disruption tolerant ad hoc networking.

# 6

# Constructing the Web of Events from raw data in the Web of Things

Yunchuan Sun[a], Hongli Yan[a], Cheng Lu[b], Rongfang Bie[b,*] and Zhangbing Zhou[c,d]

[a]*Business School, Beijing Normal University, Beijing, China*
[b]*College of Information Science and Technology, Beijing Normal University, Beijing, China*
[c]*Schoolof Information Engineering, China University of Geosciences, Beijing, China*
[d]*Computer Science Department, Institute Mines-TELECOM/TELECOM SudParis, Paris, France*

**Abstract.** An exciting paradise of data is emerging into our daily life along with the development of the Web of Things. Nowadays, volumes of heterogeneous raw data are continuously generated and captured by trillions of smart devices like sensors, smart controls, readers and other monitoring devices, while various events occur in the physical world. It is hard for users including people and smart things to master valuable information hidden in the massive data, which is more useful and understandable than raw data for users to get the crucial points for problems-solving. Thus, how to automatically and actively extract the knowledge of events and their internal links from the big data is one key challenge for the future Web of Things. This paper proposes an effective approach to extract events and their internal links from large scale data leveraging predefined event schemas in the Web of Things, which starts with grasping the critical data for useful events by filtering data with well-defined event types in the schema. A case study in the context of smart campus is presented to show the application of proposed approach for the extraction of events and their internal semantic links.

Keywords: Web of Events, Web of Things, restful, information extraction, mobile

## 1. Introduction

The Web of Things (WoT) aims at enhancing our daily life through deriving knowledge such as events (to be) happened, from the trajectories of things and people which reflect activities across the cyber world and the physical world. WoT envisions a brand-new Web that promises to provide a platform and a friendly interface for people and smart things to access the status of physical objects, where Representational State Transfer (REST) can be viewed as an architecture to support WoT [1]. Generally, WoT is enabled to sense, capture and process more data with the development of related technologies such as wireless sensor network, RFID technology and semantic technologies. Massive data are continuously generated by trillions of sensing devices in the physical world. It is impossible for people to understand and use these raw data in problem solving due to the characteristics of data such as massiveness, complexity, trivialness, dynamic and implicit semantics, which means it is rather hard for people and smart things to master valuable information hidden in the massive data.

Firstly, people are more inclined to visit intuitive, easy-to-understand and easy-to-use integrated event information rather than complex raw data. For example, it is difficult for people to understand the

---

*Corresponding author: Rongfang Bie, College of Information Science and Technology, Beijing Normal University, Beijing, China. E-mail: rfbie@bnu.edu.cn.

meaning of a set of data like (objectID: *867495012228536*, TimeStamp: 2012-03-2620:41:31, Location: (latitude=39.9661140441895, longitude=116.372123718262)) reported by the mobile device of a student, while event information like "a student has a class at 20:41:31 2012-03-26 for 53 minutes at the place of Classroom 8" is more useful and understandable for people.

Secondly, the internal semantic links among different events are more helpful and meaningful than isolated activities for people to master the situations and to solve problems. For example, a semantic link like "an event *Being absent* caused by *Playing sports*" is more useful than two isolated events "*Being absent*" and "*Playing sports*". Links between events weave the isolated events into a connected network. By using the semantic links among events, we can find out more inherent and implicit relations between events by reasoning among the existed links. Such a network of events can provide a global view of the situation for users by revealing the internal relationships between events. More useful evolving patterns can be mined from the network which can be used to predict future trends in the field. For example, we can find the inherent reasons of some absence event by the transitivity of the *causeOf* links among a series of events.

Nowadays, more and more researchers realize the importance of events to the web, and believe that events are the fundamental abstractions in studying the dynamic world. Besides, there are gradually a few works focusing on events in the web. For example [2], proposes that the evolving dynamic web would be a 'web' of interconnected events across time and space, and the paper will mainly focus on interconnections between events as they occur across space and time. Based on Twitter micro blogs, [3] studies the structural analysis of the emerging event-web; what the authors do is using occurrence and co-occurrence patterns to study Power Law in the web, and they also intend to study causality across such events in future research. Therefore, how to process these raw data efficiently and how to extract valuable knowledge from these data play key roles as well as key challenges in the development of WoT.

When extracting events in the domestic environment, the concept of event and event types should be defined, which will be further discussed in Section 3. Generally, there might be some crucial thresholds to identify an event of a certain type. For example, an event of *Having class* occurs for a student only when he is in the classroom for more than 45 minutes. In many cases, events can be identified efficiently by using a set of predefined event thresholds. Similarly, semantic links among events can also be discovered according to a list of filters on the properties of the events. Based on these ideas, an event schema which is a list of definitions on different kinds of events and different kinds of links can be worked out by domain experts; then the events and their links can be extracted by using the event schema and a network of events can be constructed; and finally we can present the events on the Web in a visualized way. Therefore, in Section 3, we introduce the concepts of event, event type, link type and event schema in WoT. Section 4 discusses the model for extracting event information which includes three layers, i.e. data collecting, event information extracting and event presenting.

At last, a case study in the smart campus has been implemented in Section 5 showing how the proposed model can be used in practice, in which our model of Event Information Extracting adopts a process procedure similar to the ECA rules (*Event-Condition-Action*). ECA rules have been widely used in various researches including semantic web. For example, [4] creates Event-Condition-Action Rule Languages in semantic web. In our model, the emergence of real data in data collecting layer(as *Event* in the ECA rules) will trigger the process of event extracting layer, where the agent judges whether the precondition (as *Condition* in the ECA rules) is satisfied; and then the critical and sensitive data meeting the conditions of the event types would be utilized to generate events (as *Action* in the ECA rules). Obviously, there is difference between our prototype and the ECA rules, for the concept of event in ECA rules differentiates from the events we want to extract by using our model.

The contributions of this paper include (i) propose an efficient approach to extract event information, especially internal semantic links between events from detailed and large scale data based on predefined event schema in WoT, and (ii) use a visualized way to present the acquired events and their internal links on WoT and to facilitate the user accessing the system.

## 2. Related work

Many efforts have been made in the fields of data management and information processing for achieving the vision of the Web/Internet of Things. Actually, researchers have done a lot of work in three different aspects of this area. A long list of challenges such as data storage, data preprocessing like cleaning and filtering, consistency checking and semantic explaining have been discussed in [5–8]. Database management system is a good start for managing sensor data. But existing database systems are somehow not suitable to be used directly for handling live sensor data [9–11]. To reduce the cost of communication, infrastructure redesign is required for data management through fully exploiting the local storage and processing capability, and a proposed architecture for managing data, i.e., *StonesDB*, is utilized for storing and processing data in sensor networks [12]. Regarding the data storage in wireless sensor networks, two approaches named as *ODS* (optimal data storage) and *NDS* (near-optimal data storage) are proposed by formalizing the storage problem into either a one-one (one producer and one consumer)or a many-many (m producers and n consumers) model [13]. Several principles are proposed for *RFID* data management, where the primary purpose is to digest the data close to the source [14]. A distributed *RFID* system is presented for efficiently cleaning, filtering and augmenting the raw data generated by tag readers [5]. Dynamic Relationship *ER* Data Model, as well as a corresponding system – Siemens *RFID* Middleware – is proposed to manage the general temporal-oriented data. Besides, this model supports *RFID* data tracking and monitoring [15]. In addition to the efforts of dealing with data from *WSNs* or *RFID* readers, some works focuson heterogeneous data from different data sources. *INDAMS* isan integrated network and data management system used for managing data in heterogeneous *WSNs* [16], while a web service based schema also aims at managing heterogeneous data in the Internet of Things [17]. A scalable lean data provision architecture based on ontology is developed to deal with the massive data in a semantic way [18]. Most of these works focus on how to store and manage the captured data rather than how to explain or extract meaningful event information.

Some works focus on the framework of data presentation and analysis. An architecture named *SemSense* is proposed for collecting real world data from a physical system of sensors and publishing the data on the Web [19]. In order to find potential and useful spatio-temporal patterns from the sensor data, a data model *STSG* (spatio-temporal sensor graphs) has been developed [20]. Those works lay emphasis on the presentation and visualization of the data but not the information extracting.

Moreover, there have been also several researches on activity recognition from the sensor data [21–32]. There are many works focusing on activity identification and recognition from data captured from various kinds of sensors, including static sensors, wearable equipment, and wrist worn *RFID* readers, etc. Environment sensors such as infrared motion detectors and magnetic door sensors have been used to gather information about complex activities such as cooking, sleeping and eating. These sensors are used in performing location-based activity recognition in indoor environments [21–23]. Ambulatory movements like walking, running, sitting, climbing and falling can be recognized from captured data through wearable sensors [24,25]. Meanwhile, many kinds of approaches have been developed to recognize activities from raw data including Naive Bayes classifiers [26,27], Hidden Markov models [28] and Decision trees [25]. Discriminative approaches including support vector machines and conditional random fields have also been effective [22].Unsupervised discovery and other recognition methods have also been introduced [29]. Transfer learning is studied under a variety of different names including learning to learn, life-long learning, knowledge transfer, inductive transfer, context-sensitive learning and meta learning for the purpose of activity recognition [30–32]. An approach is proposed to recognize multi-user activities by using Body Sensor Networks [37]. Different to the traditional data-centric methods for activity recognition, [33] introduces a knowledge-driven approach to real-time, continuous activity recognition based on multisensory data streams in smart home. Behavior patterns in active daily life are more useful than only activities, some methods to discovery activities can be used to identify behavioral patterns in observational sensor data [34,36]; another methods to mine and to monitor human activity patterns have been developed for home-based health monitoring systems [35]. All these works on activity identification and recognition lay a foundation for (i) data analysis and integration, and (ii) information extracting and explaining, in the Internet of Things. However, most of these works only focus on how to identify the isolated activities leveraging the smart contexts information, while relatively few works focus on information extracting (including both the events and their internal semantic links) from massive and heterogeneous data generated from objects in the Internet of Things.

Many semantic Web technologies such as RDF,[1] SPARQL[2] and OWL,[3] have been well developed for querying, browsing and reasoning semantic data. Some models are proposed to build semantic links, such as Linked Data which aims to build semantic links between data on the semantic web according to the internal relationship [38]. SLN schema is an emerging model in semantic network which provides a theory of normalized forms to manage and query on the complicated network [39,40]. Though those semantic tools or works are well developed or proposed, this does not eliminate the important role of information extracting in WoT where huge amounts of data are required to be timely processed. Integrating, fusing and interpreting data as well as extracting meaningful information from raw data are still big stones along the road to achieve the vision of WoT [41].

Some event models have already been proposed to extracting information based on different definitions in different domains. In the Topic Detection and Tracking (TDT) domain, an event model can be defined as a group of themes [42], while in the webpage domain, an event model (e.g. DOM model) is a series of operations related to the browser GUI such as pressing a button, clicking a hyperlink and submitting a form.[4] Some other event models are defined based on several properties such as LODE model [43] (one class and six properties) and EO model [44] (four classes and seventeen properties). Most of these definitions are based on a few characteristics. In our work, we try hard to construct the Web of Events and we propose our model on a series of comprehensive concepts such as event, event type, link type and event schema rather than a few limited characteristics, which are more suitable for the WoT scenarios where myriads of data generated all the time. So we redefine an event and identify it as an abstraction from a series of data which is more meaningful than data. And we demonstrate how an event is extracted as well as how to construct a link web in large sections.

## 3. Conceptions: Event, event type, link type, and event schema

An event is an action which occurs in the physical world and is usually accompanied by a set of correlative status changes. The monitoring devices can capture the data about the status changes when

---

[1] http://www.w3.org/RDF/.

[2] http://www.w3.org/TR/sparql11-query/.

[3] http://www.w3.org/Submission/2006/10/.

[4] http://www.w3.org/DOM/.

EventNumber: e20

EventType: Having class

Timestamp: 2012-03-2620:41:31

UserID: 200911211015

Name: Wen Liang

Location: Classroom 8

Duration: 0d,0h,53m

Fig. 1. An example event of *Having class.*

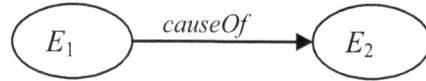

Fig. 2. An example of semantic link between events.

an event occurs in the physical world, and then transmit these data to the cyber world. The captured data should be encapsulated into units according to events specification for facilitating and promoting the understanding to the data through information extracting. In the cyber space, an event can be represented as a list of data.

**Definition 1.** An event is a tuple with the format like $E(timeDescription, ObjectsList, LocationDescription, DataList)$, where *timeDescription* records the time when the event occurs, the starting and ending time or duration, *ObjectsList* $(Oid_1, Oid_2, \ldots, Oid_n)$ is a list of objects involved in the event, *locationDescription* records the location where the event occurs, and *DataList* is a list of data describing the event-related status or interrelations of the objects in *ObjectsList*. In the Internet/Web of Things, the *DataList* might be a summary of the raw data captured across time and space.

For example, we can use $e20$ (*Having class, 2012-03-2620:41:31,Classroom 8, Wen Liang, 0d0h53m*) to depict an event "*Wen Liang has a class at 20:41:31 2012-03-26 for 53 minutes at the place of Classroom 8*" where the data in the list is filtered from the raw data. Event information like this is more intuitive, easy-to-understand and easy-to-use than raw data for users or intelligent systems.

Often, not all data sensed from the physical world is useful to restore the event information. Moreover, not all events are useful for users to grasp the situation or to solve problems. An intuitional idea is to extract useful event information from the detailed raw data based on well-defined event types provided by the domain experts or intelligent mining tools.

There are various types of events in our daily life, and for events with the same type, similar solutions can be adopted due to their common characteristics. In the cyber space, an event type can be defined as a model to formulate the data of the event instances with the same type.

**Definition 2.** An event type *ET* is a 2-tuple (*Precondition, EventDescription*), where *Precondition* is a logic expression $(P_{11} \wedge P_{12} \wedge \ldots \wedge P_{1k_1}) \vee (P_{21} \wedge P_{22} \wedge \ldots \wedge P_{2k_2}) \vee \ldots \vee (P_{n1} \wedge P_{n2} \wedge \ldots \wedge P_{1k_n})$, each part of which is a condition to the event. *EventDescription* is a data record consisted of name and descriptions of the event like (*timestamp, ObjectsList, LocationDescription, DataList*) as defined in Definition 1.

The event types are used to regularize the events which can be mined from the massive and heterogeneous data. When the captured data meets the condition *Precondition* in Definition 2, an event of the defined *ET* type is confirmed, and the event information *EventDescription* would be posted on the web.

For example, the event type of *Having class* can be defined as follows:

The Event of *Havingclass*: $((duration \geqslant 45 \text{ mins}) \wedge (locationName = \text{Classroom})$, $(start\_time, end\_$

*time, duration, location, Having class*))

While an mobile device reports a measurement (latitude = 39.9661140441895, longitude = 116.372123718262) of a person (with a *objectID = "867495012228536"*) to the system at 2013-03-26 20:41:31, the agent can search some necessary information of the people from the Resource Repository according to *objectID* like *UserID* = 200911211015, *Location* = Classroom 8 and *name* = Wen Liang, then determines whether the precondition is satisfied for the event type of *Having class*. The agent would name the event and post it on the web in the form shown in Fig. 1.

Teachers or other users can access the event information through the web and understand the event: *"Wen Liang has a class at 20:41:31 2012-03-26 for 53 minutes at the place of Classroom 8"*.

**Definition 3.** A link between two events is a 3-tuple *L(startEvent, endEvent, factor)*, which means that there is a *factor* relation linking *startEvent* to *endEvent*.

For example, *time sequence* (like *direct-succeeding*) and *causeOf* are two general kinds of links between smart campus events. (*E*1, *E*2, *causeOf*) means that an event *E*1 *causes* another event *E*2, which can be shown in Fig. 2.

Links between events weave the isolated events into a connected network which might be more useful than a set of isolated events and provides a global view of the situation for users by letting them grasp the internal relationships between events more easily. More useful evolving patterns can be mined from the network which can be used to predict future trends in the field, and potential relations between events can be deduced based on the existing links. For example, we can find the inherent reasons of some illness event by the transitivity of the *causeOf* links among a series of events.

Indeed, there are various kinds of link types with different factors. The link types can be determined by a pair of event types, that is, between a pair of event types there are some certain link types with certain factors.

**Definition 4.** A link type *LT* between event types can be represented as *LT(startEventType, endEventType)*, which means there is a link type from event type *startEventType* to event type *endEventType*. The link factor between two events can be defined by some certain inherent properties of the events.

A set of link types can be defined according to the requirements of a certain domain. For two certain events, the semantic links between them can be determined by their inherent properties. For example, a *direct-succeeding* link between an event of *Having class* and another of *Having lunch* is determined by the end time of the lunch event and the start time of the class event. The filters to determine the semantic links between events can be worked out according to the actual applications. Even more, there might be a series of reasoning rules among these link types which can be used to deduce the potential internal relationships implied in the existing links. For example, the link type *causeOf* has the character of transitivity, i.e., *causeOf\*causeOf → causeOf*.

For a given application, experts can define a set of useful event types, and a set of link types as well as a set of reasoning rules based on the link types. These well-defined event types, link types and the reasoning rules construct a domain- or application-dependent event schema. Indeed, an event schema is a domain-dependent knowledge base to differentiate the critical and sensitive information from the massive and heterogeneous data.

**Definition 5.** An event schema has 3 components: (i) a set of event types $\{ET_1, ET_2, \ldots, ET_n\}$ where each $ET_i$ is an event type, (ii) a set of link types $\{LT_1, LT_2, \ldots, LT_m\}$ where each $LT_k$ is defined between a pair of $(ET_i, ET_j)$, and (iii) a set of reasoning rules $(r_1, r_2, \ldots, r_l)$ where each $r$ is in the form of $LT_i*LT_j \to LT_k$.

Event schemas may vary depending on different application scenarios and play the most important role during the extracting process. Traditionally, the event schema should be defined by the domain experts or mined from a large scale of history information.

## 4. A model for event information extracting

### 4.1. Framework of the Web of Events

Figure 3 shows the framework of the Web of Events. In the Internet of Things, trillions of smart things are distributed in different types of application scenarios such as e-Health, logistics and smart homes. The event extracting agents in these application scenarios extract various meaningful events constantly from the captured large scale raw data according to the local event schemas provided by domain experts. The derived events would be transmitted to the Web of Events. The Web of Events can organize the events in a multi-classification way and be capable of integrating the events from different scenarios seamlessly for specific purposes. The Web users or intelligent agents can access to the events easily through REST related technologies like HTTP or other mobile protocols.

In this framework, there are two key challenges: (i) How to extract the event information from the raw data. The event schema consisting of a series of event types plays the most important role during the extracting process. Indeed, the event schema is a base of knowledge derived from the domain experts or intelligent knowledge mining tools. (ii) How to organize and integrate events from different application scenarios. These events would be endowed with enriched semantics. Besides, the various semantic relationships among the events would be helpful to weave an event network with semantic links which would be more useful for implementing the intellectualization of the Internet of Things. This paper focuses on the former challenge: *the event extracting*.

### 4.2. Model for event information extracting

The process of extracting event information consists of three phases: (i) collecting raw data by sensing devices, (ii) parsing the captured data and extracting event information, and (iii) organizing and presenting event information on the web. Therefore, the model for event information extraction is 3-layered architecture as shown in Fig. 4. The data collecting layer aims at capturing raw data from different kinds of application scenarios with the support of smart sensors and transmitting the data to the upper layer with some specific formulation; the event extracting layer encapsulates raw data with the basic information of related objects stored in the system and extracting the event information with the help of the event schema; and the aims of the presenting layer lie in two aspects: 1) organizing and integrating the events in a way of category and semantics; 2) providing a platform for Web users and intelligent agents to access the events.

While the data flow continuously pipes into the system, the most important point is to extract useful information by encapsulating the interrelated data in the event extracting layer. The details about the model are discussed as follows.

#### 4.2.1. Data collecting layer

In the data collecting layer, sensors, RFID readers or other monitoring devices collect data from the physical world in time, and transmit the collected data to *Data Agent* in the event extracting layer with

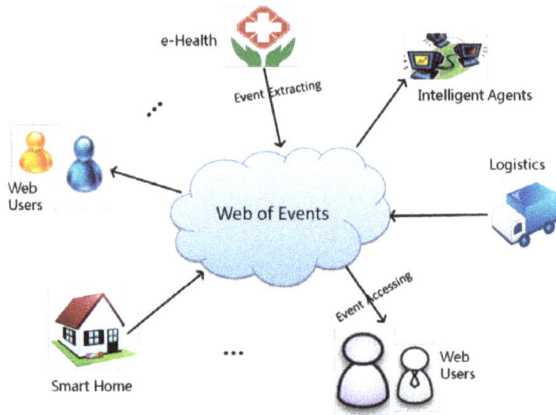

Fig. 3. A framework of the Web of Events.

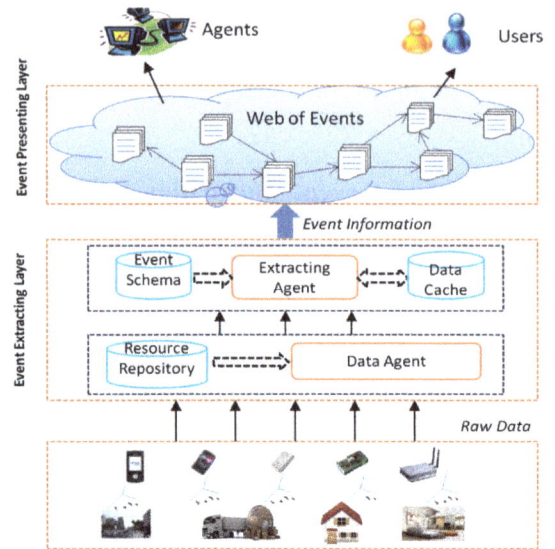

Fig. 4. The model for eventinformation extracting from raw data in WoT.

specific formats upon device types. The timestamp, target object ID and captured content (such as *temperature*, *humidity*, *location* or other measurements) are required in each item of the data. Each monitoring device is endowed specified functions by the manufacturers and the engineers of the applications.

The collected data formats depend on the monitoring devices in specific applications. In most cases, the data is massive and heterogeneous due to the various kinds of different devices and the long-term continuously detection. Critical and sensitive information about useful events is drowned in these trivial and complex data.

### 4.2.2. Event extracting layer

The event extracting layer aims at extracting useful event information from the massive and heterogeneous data, with the support of five components: *Resource Repository*, *Data Agent*, *Data Cache*, *Extracting Agent* and *Event Schema*.

1. *Resource Repository* stores the primary inherent information like name, manufacturer, timestamp and other perpetual attributes of each monitored objects in the application. The perpetual information might support the event extracting process. In practice, the resource repository might be put on the web or in the cloud store service.

2. *Data Agent* is responsible for data preprocessing including analyzing the source, target object, explaining semantic content of the raw data, searching out the basic information of the object from the repository and shunting the well-processed data to the extracting agent.

In the Internet of Things, each target object in the physical world would be identified by a unique ID. Usually, each record in raw data from the data collecting layer can be transformed into the format like {*Oid*, *DataType*, *Value*, *Timestamp*}, where *Oid* represents the target object, *DataType* is the type of the value which specifies the expected purpose of the monitored data, *Value* is the monitored value of the target object, and *Timestamp* records the time when the data is generated. In line with the unique *Oid*, each record of the data can be merged with the corresponding inherent information of the object stored in Resource Repository, and then the data record could be encapsulated into a specific format like (*Oid*,

*DataType*, *Value*, *Timestamp*, *InformationList*), where *InformationList* is a list of inherent information of the target which is useful to the generated events. Then the encapsulated data would be transmitted to the extracting agent for further processing.

The event schema stores information of all possible event types and link types in the application each of which includes a precondition.

1. *Data Cache* stores history data from data agent which might be used in the following processing. The operating principles for the extracting agent are listed as follows.
2. Once some data records arrive, the *Extracting Agent* first make sure the relevant event types which the data might trigger. There might be not only one event type related to the coming data. For each possible event type, the agent judges whether the precondition is satisfied, sometimes with the support of the data cache which can provide the necessary history data. During this judgment, the agent produces a new event with the provided data if the data meets the precondition and transmits the event information to the event presenting layer. Otherwise, the data would be put into *Data Cache* if they might be useful in the following analysis. Or the data would be thrown away.
3. *Event Schema*. Obviously, only the critical and sensitive data meeting the conditions of the event types would be utilized to generate events. Most data would be thrown away after the extraction. The semantics of the generated events is more easy-to-understand than the raw data. The process of the event extracting can somehow be viewed as a process of information discovery and data explanations.

Once the events have been extracted, we can check the internal links between each pair of events according to the predefined link types and their filters. Then, a network of event is generated with a set of semantic links among events.

### 4.2.3. Event presenting layer

The event presenting layer organizes the events according to their types, timestamp and the internal links among them or other information, and presents the event information in intuitive ways to users through web tools like HTTP or to intelligent systems with general interfaces for the purpose of decision making or problem solving automatically and autonomously.

## 5. Case study – web of smart campus events

The advent of the Internet/Web of Things enables a lot of new application scenarios in business and daily life. This section introduces a case study to show the applicability of the proposed model presented above.

### 5.1. Scenario

A smart campus is an intelligent agent that perceives the state of resident and the physical environments using various kinds of devices like mobile phones, sensors and readers. These devices can capture massive concrete data about individual's activities, environment settings and inhabitants' characteristics. In this section, we'd like to illustrate how to construct the Web of Events through the proposed models.

On the one hand, to better verify our model's performance in dealing with heterogeneous data, we build a student attendance system in our laboratory. One reader as well as eight antennas is installed and these antennas are deployed in eight regions of two rooms. 14 RFID tags are assigned to 14 students respectively. On the other hand, we develop an android program to collect two person's daily events manually and their GPS every five minutes automatically. Therefore, we collect two diverse datasets from these two different scenarios for about one week.

Table 1
Event types in the mobile scenario

| Event type | Conditions and stored information | Descriptions |
|---|---|---|
| *Having breakfast* | ((6:30<*startTime*<9:30)∧(*duration*>=5 mins) ∧ (*locationName* = Canteen), (*start_time, end_time, duration, location, Having breakfast*)) | *Having breakfast* is the type of events which occurs in the canteen (obtained from GPS, similarly hereinafter) lasting for no less than 5 mins and the start time is between 6:30 and 9:00 (based on school timetable). |
| *Having lunch* | ((10:30<*startTime*<13:00)∧(*duration*> =20 mins)∧ (*locationName*= Canteen), (*start_time, end_time, duration, location, Having lunch*)) | *Having lunch* is the type of events which occur in the canteen lasting for no less than 20 mins and the start time is between 10:30 and 13:00 (based on school timetable). |
| *Having supper* | ((16:30<*startTime*<18:30)∧(*duration*> =20 mins)∧ (*locationName*= Canteen), (*start_time, end_time, duration, location, Having supper*)) | *Having supper* is the type of events which occur in the canteen lasting for no less than 20 mins and the start time is between 16:30 and 18:30 (based on school timetable). |
| *Playing sports* | ((*duration*>=30 mins)∧(*locationName*=Playground), (*start_time, end_time,duration, location, Playing sports*)) | *Playing sports* is the type of events which occur in the playground lasting for no less than 30 mins. |
| *Having class* | ((*duration*> =45 mins)∧ (*locationName*= Classroom), (*start_time, end_time,duration, location, Having class*)) | *Having class* is the type of events which occurred in the classroom lasting for no less than 45 mins (one class). |
| *Studying* | ((*duration*> =30 mins)∧(*locationName*= Library), (*start_time, end_time,duration, location, Self-Studying*)) | *Self-Studying* is the type of events which occur in the library lasting for no less than 30 mins. |
| *Working* | ((*duration*> =30 mins)∧(*locationName*= Laboratory), (*start_time, end_time,duration, location, Working*)) | *Working* is the type of events which occur in the laboratory lasting for no less than 30 mins. |
| *Being in the dorm* | ((*duration*> =30 mins)∧(*locationName*= Dormitory), (*start_time, end_time,duration, location, Being in the dorm*)) | *Working/Relaxing* is the type of events which occur in the dormitory lasting for no less than 30 mins. |

## 5.2. Event schema for smart campus

Our schema depends on the real dataset. Firstly, we work out the Event Schema of Smart Campus which involves in a series of event types: *Having breakfast, Having lunch, Having Supper, Entertainment, Discussion, Playing sports, Having class, Studying, Working, Being in the dorm, Leave temporarily, Being absent, Leaving early* and *Being late.* The last four event types are suitable for detecting students' attendance events, while the others are for daily events in campus.

Time range for each event type is necessary to identify an event. But in different scenarios, time ranges for each event type could be different. Thus, we define time range based on school timetable or laboratory timetable. The event types in the mobile scenario and RFID scenario are listed in Tables 1 and 2 respectively.

All event types are defined in advance, and the conditions of event types except *Entertainment* and *Discussion* are shown in Table 1. *Entertainment* and *Discussion* used in the android program are defined by us too, but a concrete event is selected by user according to specific circumstance. As for automatically collected data, we gain event types according to users' submitted information and Table 1 (if submitted information is missing).

Then we can work out possible link types among the above event types. Herein, for the consideration of simplification, only five link types (*d-succeeding, co-occur, overlap, sameTypeOf* and *causeOf*) between any two event types are defined according to the actual data as shown in Table 3, where we also

Table 2
Event types in the RFID scenario

| Event type | Conditions and stored information | Descriptions |
|---|---|---|
| *Leave temporarily* | $((8{:}30{<}startTime,endTime{<}11{:}30)\vee(14{:}00{<}startTime, endTime{<}16{:}40))\wedge(duration{>}=10$ mins$)\wedge(locationName =$Out of Laboratory$), (start\_time, end\_time,duration, location, Leave\ temporarily))$ | *Leave temporarily* is the type of events which occur out of the laboratory (monitored by a RFID reader, similarly hereinafter) lasting for no less than 10 mins, and both the start time and the end time are within the range of 8:30 ∼ 11:30 in the morning or 14:00 ∼ 16:40 in the afternoon (based on laboratory timetable). |
| *Being absent* | $((startTime{<}8{:}30)\wedge(endTime{>}11{:}30)\vee(startTime{<}14{:}00)\wedge (endTime{>}16{:}30))\wedge(locationName=$ Out of Laboratory$), (start\_time, end\_time, duration, location, Being\ absent))$ | *Being absent* is the type of events which occur out of the laboratory and both the start time and the end time are not in the range of 8:30 ∼ 11:30 and 14:00 ∼ 16:40 (based on laboratory timetable). |
| *Leaving early* | $((startTime{<}11{:}30{<}endTime)\vee(startTime{<}16{:}40{<}endTime)) \wedge(duration{>}=30$ mins$)\wedge(locationName=$Out of Laboratory$), (start\_time, end\_time, duration, location, Leave\ temporarily))$ | *Leave temporarily* is the type of events which occur out of the laboratory (monitored by a RFID reader) lasting for no less than 30 mins and the start time is earlier than 11:30 in the morning or 16:40 in the afternoon (based on laboratory timetable). |
| *Being late* | $((startTime{<}8{:}30{<}endTime)\vee(startTime{<}14{:}00{<}end Time))\wedge(locationName=$Out of Laboratory$), (start\_time, end\_time, duration, location, Being\ late))$ | *Being late* is the type of events which occur out of the laboratory (monitored by a RFID reader) and the end time is later than 8:30 in the morning or 14:00 in the afternoon (based on laboratory timetable). |

list the descriptions of specific link type and its corresponding format. Meanwhile, we pose the judgment condition of specific link type, so that by using the conditions links can be generated by system automatically.

### 5.3. Architecture of the prototype

To verify the event extraction model, we have built a prototype using Tomcat as the Web server and MySQL5.5 as the database. Figure 5 shows the overall event extraction process in a simplified Smart Campus application.

In the data collection layer, we gain the captured RFID data flow from the attendance system and mobile data from android procedure ①. The attendance system captures RFID binary data all the time and transforms data into the format of {*Oid, DataType, Activeid, Timestamp*}, where *Oid* is RFID tag to identify the target person, *DataType* is used to describe that the captured data is from RFID source, *Activeid* is detected antenna ID when a person enters corresponding region, and *Timestamp* indicates the measured time. The preprocessed RFID data will be sent to the extracting layer in JSON format, which maybe more adapted to devices with limited capabilities [45].

The format of mobile data recorded by person manually is {*Oid, DataType, Etype, Title, Note, Latitude, Longitude,Timestamp*}, where *Oid* is the serial number of the target mobile, *DataType* is used to describe that the captured data is from mobile source, *Etype* is the general event type defined by users, *Title* and *Note* is event title and content of an event recorded by a user, *Latitude* and *Longitude* is GPS information recorded automatically, and *Timestamp* indicates the recorded time. Besides, a person's GPS information will be captured every five minutes automatically to excavate those unrecorded events man-

Table 3
Link types between event types in smart campus

| Link type | Format | Conditions | Descriptions |
|-----------|--------|------------|--------------|
| D-succeeding | $(e_1, e_2,$ d-succeeding) | $((e_1.end\_time < e_2.start\_time) \wedge$ $(e_2.start\_time < e.start\_time)$, where $e$ is any event such that $e.start\_time > e_1.end\_time$. | $e_2$ is a directly succeeded event after $e_1$ ends. |
| Co-occur | $(e_1, e_2, co\text{-}occur)$ | $((e_1.start\_time < e_2.start\_time) \wedge$ $(e_1.end\_time > e_2.end\_time) \vee$ $((e_2.start\_time < e_1.start\_time) \wedge$ $e_2.end\_time > e_1.end\_time)$ | $e_1$ occurs in the period of $e_1$, vice versa. |
| Overlap | $(e_1, e_2, overlap)$ | $((e_1.start\_time < e_2.start\_time) \wedge$ $(e_1.end\_time > e_2.start\_time) \wedge$ $(e_1.end\_time < e_2.end\_time) \wedge$ $(e_2.start\_time < e.start\_time))$, where $e$ is any event such that $((e.start\_time > e_1.start\_time) \wedge$ $(e.start\_time < e_1.end\_time) \wedge$ $(e.end\_time > e_1.end\_time))$ | There is an overlap period between the periods of $e_1$ and $e_2$. |
| SameTypeOf | $(e_1, e_2,$ sameTypeOf) | $(e_1.type = e_2.type)$ | $e_1$ and $e_2$ are with the same event type. |
| CauseOf | $(e_1, e_2, causeOf)$ | $e_1.start\_time < e_2.start\_time$, and other conditions should be defined according to the event types of $e_1$ and $e_2$. | $e_2$ is caused by $e_1$. For example, A person is absent because he goes to play sports. |

ually. The format of GPS information is {*Oid, DataType, Latitude, Longitude, Timestamp*}. Both kinds of mobile data will be sent in JSON format.

In the event extracting layer, the data agent process ② will first search out the basic information of each data record according to relative unique *Oid* from a person information Table ③. Information of latitude and longitude will be converted into specific location. Then each data record will be encapsulated into the format like {*Oid, DataType, Value, Timestamp, Location* ... }, and the well processed data list will be shunted to the extracting agent process ④.

The extracting agent process undertakes the most important work of event extracting. Firstly, it will find out the relative event types according to the data type of each data record from a mapping Table ⑤. Then, for each event type of each data record, it will call two main sub-process modules – the data caching module ⑥ and the event generating module ⑦. Each event type has a data caching table. The data caching is responsible for data caching and labeling the changing state of data as well as its starting and ending time from which we can calculate the duration time of an event. The event generating module decides whether and what event should be generated according to the changing state of data. Both modules are sharing a group of event types ⑧.

In the event presenting layer, the generated events in the event table will be presented on web pages with the help of corresponding processes ⑨. Figure 8 shows the final pages to users, where the left web page shows the event list. These events are archived according to dates and can be accessed by event name, event type, timestamp, objected or name. The detailed event information of one specific event is showed on the right web page.

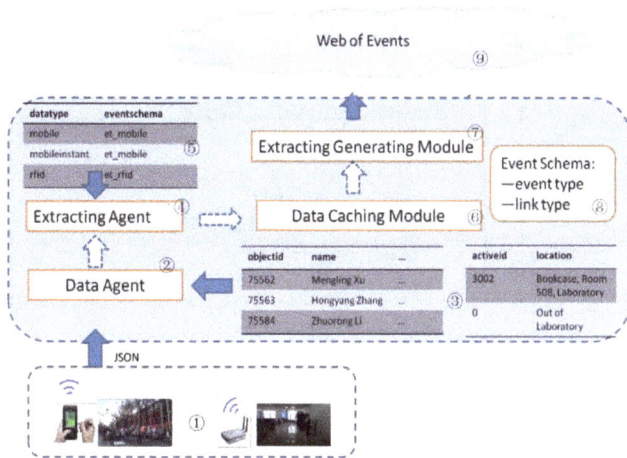

Fig. 5. Architecture of the prototype.

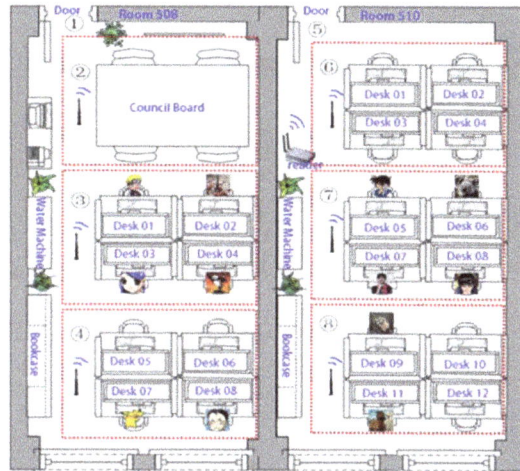

Fig. 6. The student attendance system.

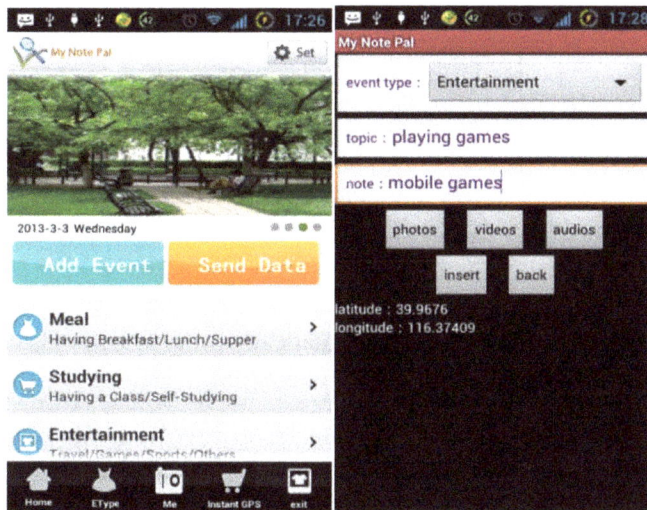

Fig. 7. Using android procedure for collecting data.

Table 4
Event counts of nine event types

| EventType | Counts |
| --- | --- |
| Being absent | 3 |
| Being late | 1 |
| Being in the dorm | 11 |
| Discussion | 3 |
| Entertainment | 27 |
| Having class | 18 |
| Having breakfast | 9 |
| Having lunch | 11 |
| Having supper | 14 |
| Leaving early | 1 |
| Playing sports | 1 |
| Studying | 23 |
| Working | 5 |
| Total | 127 |

## 5.4. Implementation and presentation

### 5.4.1. Data collecting

We build a student attendance system in our laboratory (See Fig. 6). The system has a reader installed in Room 510, eight antennas installed in eight regions (1–8) of Room 508 and 510, and 14 active RFID tags assigned to 14 persons. When a person with a RFID tag is entering region 8, the reader will detect the activated RFID tag and send data (incl. tag ID and antenna ID) to the host computer in the binary format. The reader can detect the person's location according to the corresponding antenna (active id: 3006). Then the binary data will be processed and sent to the extracting server in a JSON format, like: [{"oid":"75592","datatype":"rfid","activeid":"3008","timestamp":"2013-03-07 16:54:30"}].

We also develop an android program for collecting persons' daily events (See Fig. 7). One can record

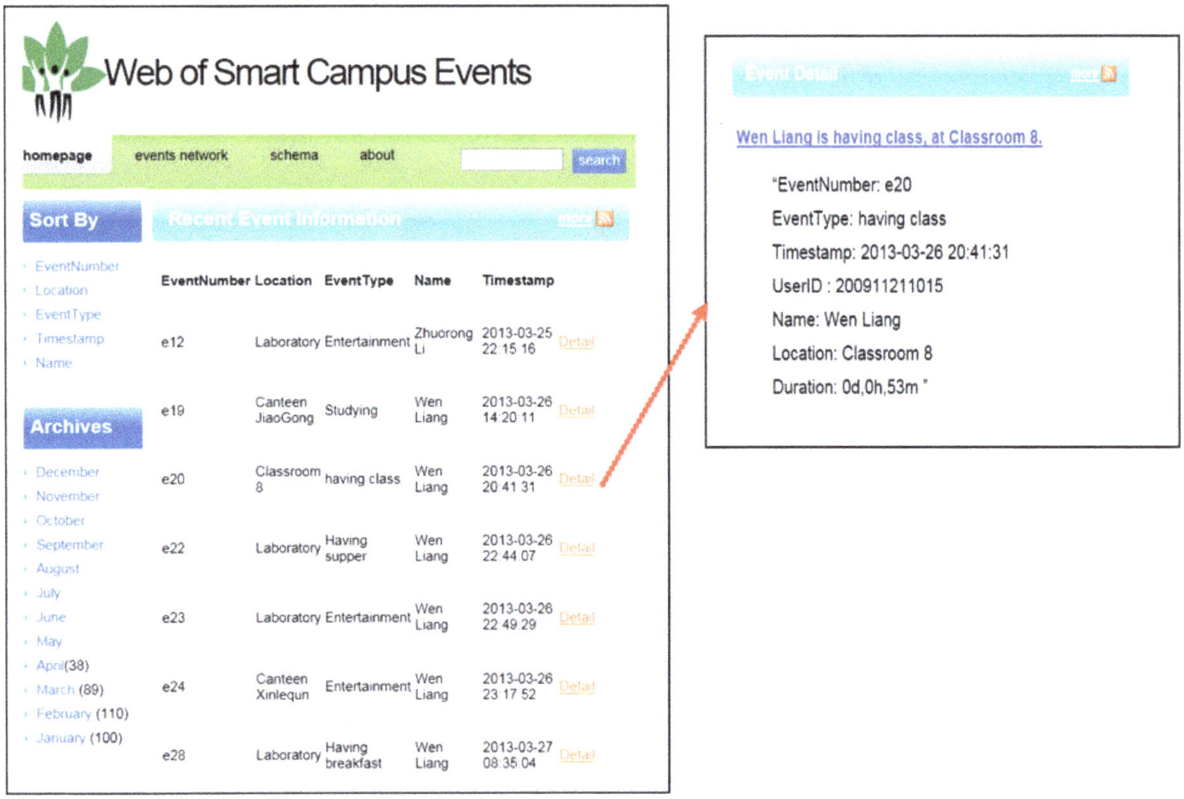

Fig. 8. Interface for smart campus event presentation.

his daily events manually, while his GPS information will be captured automatically every five minutes. These two kinds of mobile data will be sent in JSON format automatically, like:

[{"oid":"352274016686995","datatype":"mobile_noteinstant","timestamp":"2013-04-02 17:16:33", "longitude":116.37403869628906,"latitude": 39.96767807006836}]

[{"oid":"352274016686995","datatype":"mobile_note","etype":"Entertainment","topic":" playing games", "note":"mobile games", "timestamp":"2013-04-02 17:28:26", "longitude": 116.3740925781 25, "latitude":39.96764155883789}]

To make the data collecting layer more flexible and more suitable, we turn the physical devices (RFID tag or mobile) into RESTful resources and offer a uniformed REST API to access these devices. Request the resources will be formulated using a standard URL. For example, we can contact the root URL with the verb POST to add a mobile node: http://ireg.bnu.edu.cn/ WoEServer /datacollect/mobile/addnode/.

We can delete the mobile node whose oid is 352274016686995 by contacting the URL:

http:// ireg.bnu.edu.cn/ WoEServer /datacollect/ mobile /deletenode/352274016686995

The following URL are used to communicate mobile data and GPS information with the server respectively:

http:// ireg.bnu.edu.cn/WoEServer/datacollect/ mobile /data

http:// ireg.bnu.edu.cn/WoEServer/datacollect/ mobile /instantdata.

Table 5
Link counts of five link types

| Link Type | Counts |
| --- | --- |
| d-succeding | 100 |
| co-occur | 43 |
| overlap | 3 |
| sameTypeOf | 114 |
| causeOf | 0 |
| Total | 260 |

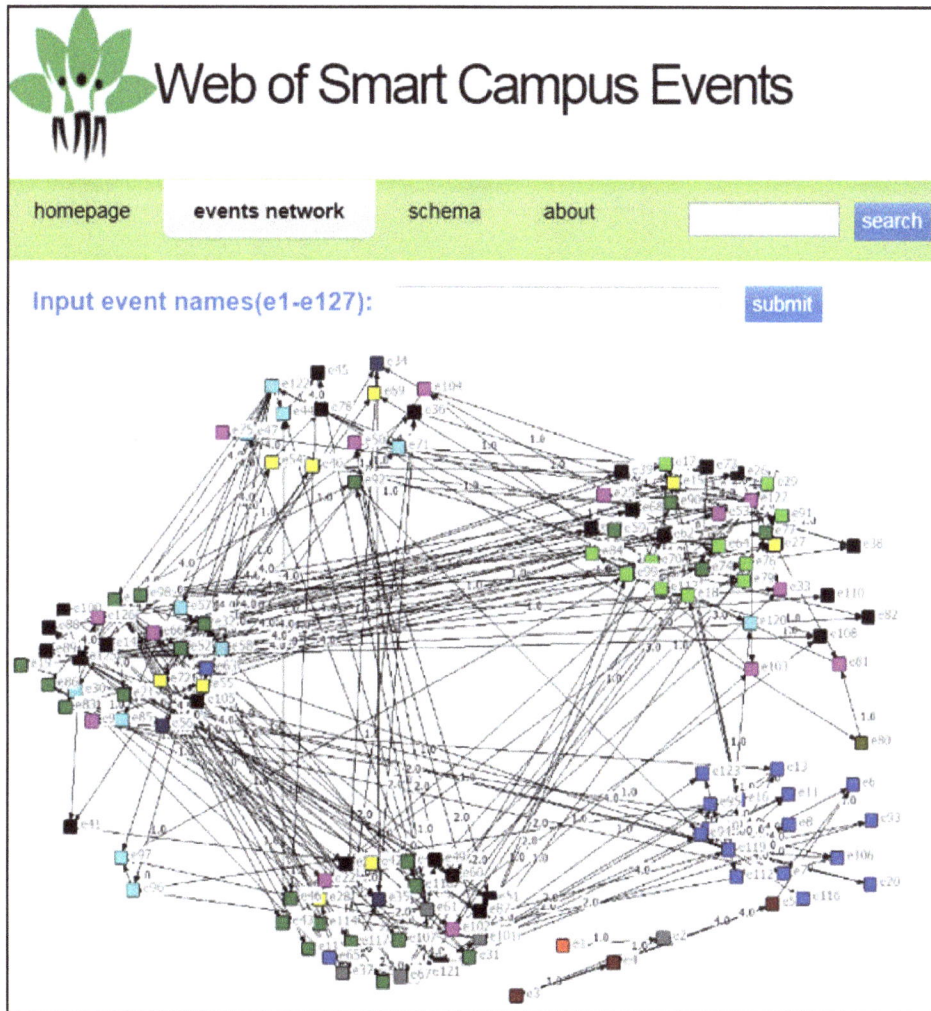

Fig. 9. Interface for smart campus event link presentation.

### 5.4.2. Event extracting and presentation

At the beginning of the extracting layer, Data Agent will find out specific location information of the submitted data. For RFID data, we can gain location according to the active id of each data record from a mapping table. For mobile data, the indoor error of latitude and longitude is greater than outdoor error as GPS signal is unavailable in the room. Through experiment we find that the error is stable in

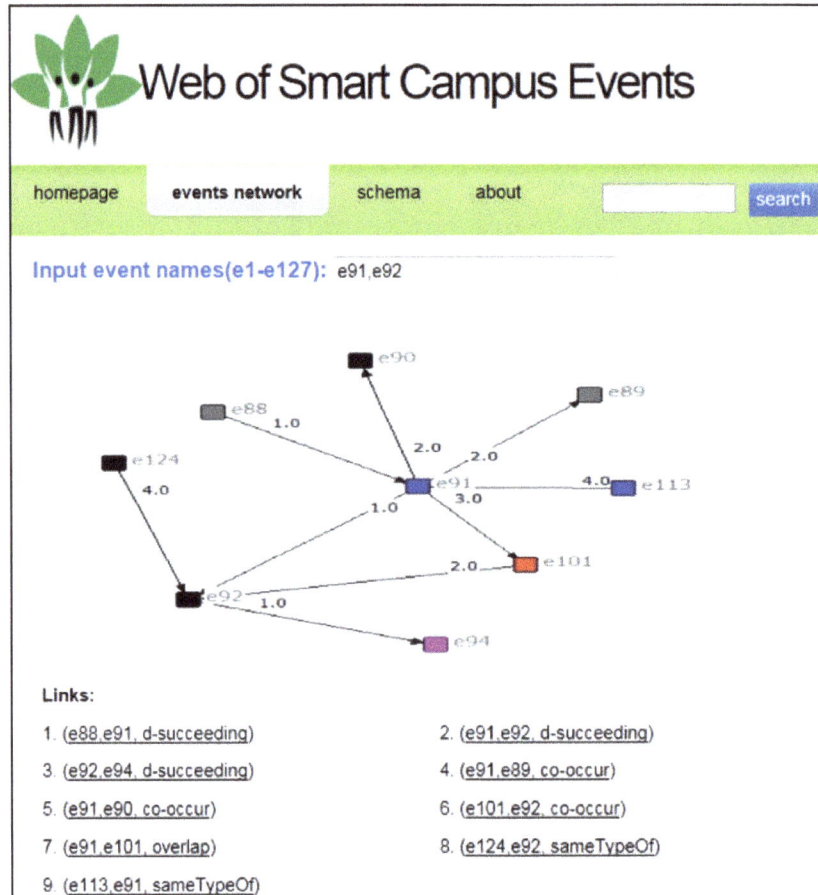

Fig. 10. Query results: links connected to two selected events e91 and e92.

the same location. So we randomly select a group of points (with latitude and longitude) to represent the corresponding location. By calculating whether a target point belongs to the represented points, we can find the target location. Then the data will be converted into the format of (*oid*, *datatype*, *location*, *timestamp*, *duration*, *name*).

  {"oid":"352274016686995","datatype":"mobile_note","location":"Laboratory","timestamp": "2013-04-02 17:28:26", "name": "Zhuorong Li"}

We have developed a procedure to extract the event information according to the above defined event types. Total 127 events of13 types have been extracted from the dataset as shown in Table 4, and 260 links between events of 5 different link types have been retrieved according to the predefined schemas as shown in Table 5. We also developed a tool by using JSP language to present the event information on the Web as shown in Fig. 8.

### 5.4.3. Link extracting and presentation

We have developed a java procedure to extract the five kinds of link information according to the above defined link types. To well represent link information of the extracted events, we use asocial

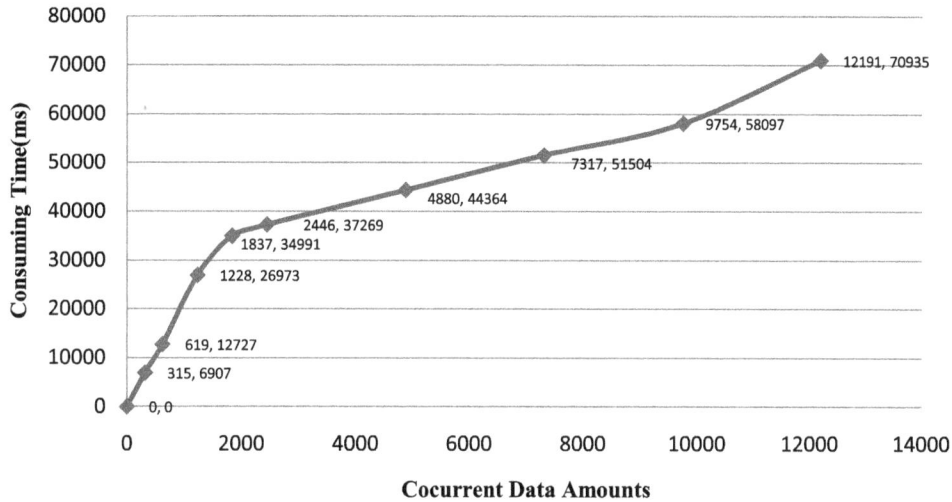

Fig. 11. Consuming time of concurrent data for event extracting layer.

network analysis tool NetDraw[5] for drawing graphs of link information, as shown in Fig. 9, where each node represents an event and each edge represents a semantic link with a link type between nodes. The number on a link reflects corresponding link type.[6] Different colors of nodes reflect different event types. There are almost five groups of nodes in Fig. 9, which reflect the most frequent events happen in five locations. For example, the right groups in color blue reflect having classes in the classroom. While some events are selected, a set of links connected to the selected events can be presented in a way of visualized network or text format as shown in Fig. 10. The developed tool can also provide interfaces to smart devices to support the intelligent decision making systems. Our system offers a uniform RESTful interface. Request the resources will be simply formulated using a standard URL. For example, we can visit some events related to someone with userID 352274016686995 by contacting a URL with the verb GET: http:// ireg.bnu.edu.cn/ eventpresent /352274016686995.

### 5.4.4. Evaluation

The processing time mainly includes three parts: data transmission time from the physical world to the event extracting layer, data processing time and event extracting time in the event extracting layer, event presenting and link presenting time in the event presenting layer. Data transmission time depends mainly on many external factors such as wireless network and network traffic. Event presenting is not in real-time. So as for a system dealing with real-time data, the key important time performance we should consider is time spending on the event extracting layer. We build our event extracting program on a desktop computer equipped with a 4 G memory, Intel(R) Core(TM) i5-2400 CPU and Windows 7 operating system. Time taken to deal with a mobile data or a RFID data at some time is less than 50 ms.

But in an instant physical environment, myriads of data may be transmitted to the extracting layer at the same time. To simulate this situation, we retain the collected data[7] for one week as much as 12191, including 100 notes recorded by two persons, 689 GPS information recorded by mobile system

---

[5]http://www.analytictech.com/Netdraw/netdraw.htm.

[6]1.00 stands for *d-succeeding*; 2.00 stands for *co-occur*; 3.00stands for *overlap*; 4.00 stands for *sameTypeOf* ; 5.00 stands for *causeOf*.

[7]Notice that in the real scenario, raw data will not be retained except the data cache model.

automatically, and 11402 RFID data. We store them separately into three tables and select9 groups of points as shown in Fig. 11, where we conclude that as concurrent data increases consuming time increases. Our model works well even when the concurrent data is large. For example, consuming time is about one minute when concurrent data amount is 9754. Concurrent data increases almost 1000 (from 2446 to 12191) while consuming time increases less than one minutes.

## 6. Conclusion and discussion

As tremendous data emerging into our life along with the development of the Web of Things, people are more willing to master the situations based on event information than raw data. How to extract the event information and internal links between events by integrating heterogeneous data in the Internet of Things is a key challenge. This paper first introduces the conceptions of *event, event type, link types* and *event schema*. Then, we propose a framework of the Web of Events to present the events on the Web of Things for the convenience of access to users and intelligent systems. A 3-layered model is proposed for extracting events and their internal relations from large-scale data set based on predefined event schema in a certain domain. The proposed model is an easy-to-understand and easy-to-use tool for filtering useless or meaningless data, so that people can analyze and solve problems by grasping the key events quickly and intuitively. A case study on smart campus is developed to show how to use this model to extract event information (including events and their internal semantic links) from the physical world in an efficient and effective way. The advantages of our approach are as follows:

- The proposed model provides a uniform framework to facilitate the decision-making of people or smart agents by integrating heterogeneous data into understandable event information. Heterogeneous data from different kinds of sensing devices or readers in different places can be used in the event extracting process. Users access the event information extracted from the raw data in a friendly and understandable way.
- The model provides a way to compress massive data generated with the explosive increasing of sensors by neglecting the redundant data. The massive data in the Internet of Things bring several challenges such as how to store, how to manage, how to query, how to process and how to use. The proposed model throws off the redundant data and only reserves the useful and crucial part.
- The Web of Events provides a platform to organize and represent the events extracted from physical world in a friendly and intuitional way. It also provides an efficient tool for further analysis on the interrelated events, especially for the behavior pattern analysis. There might be various kinds of interrelations among the extracted events like sequence (time), cause-effect and consist-of. We can retrieve more potential and implied relations through reasoning on these rich relations. In addition, valuable knowledge among a number of events and their internal links can be mined from raw data and the knowledge would be useful to patterns discovering.

Location is very important for event generating. In our case study, we randomly select a group of points (with latitude and longitude) to represent the corresponding location. There may be other ways to gain location information such as data mining. But which way is better? This problem needs to be resolved in the future efforts. There may be other ways to extract event information such as data mining and ontology. In future, we will try to find other more appropriate ways to extract events as well as improve the performance of our system. In this work, we construct the Web of Events which is simple but flexible enough. In future, we will focus on how to design a reasonable and efficient semantic model of event-event, event-person or person-person based on the emerging technologies of semantic sensor networks domain and continue to explore how to solve problems or to find useful potential patterns with the support of the Web of Events.

## Acknowledgments

This research is supported by National Natural Science Foundation of China (61171014) and the Fundamental Research Funds for the Central Universities.

## References

[1] D. Guinard and V. Trifa, Towards the Web of Things: Web mashups for embedded devices, in *Proceeding of WWW* (*International World Wide Web Conferences*), Madrid, Spain, 2009.

[2] R. Jain, Eventweb: Developing a human-centered computing system, *IEEE Computer* **41**(2) (2008), 42–50.

[3] V.K. Singh and R. Jain, Structural analysis of the emerging event-web, in Proceedings of the 19th international conference on World wide web. ACM, 2010, pp. 1183–1184.

[4] G. Papamarkos, A. Poulovassilis and P.T. Wood, Event-condition-action rule languages for the semantic web, Workshop on Semantic Web and Databases, 2003, pp. 309–327.

[5] S.S. Chawathe, V. Krishnamurthy, S. Ramachandran and S. Sarma, Managing RFID data, in *Proceedings of the 30th VLDB Conference*, 2004, pp. 1189–1195.

[6] J. Liu and A. Terzis, Sensing data centres for energy efficiency, *Philosophical Transactions of the Royal Society A: Mathematical, Physical and Engineering Sciences* **370** (January 2012), 136–157.

[7] S.R. Jeffery, M.G. Garofalakis and M.J. Franklin, Adaptive cleaning for RFID data streams, in *Proceedings of the 32nd international conference on Very large data bases (VLDB)*, 2006, pp. 163–174.

[8] S.R. Jeffery, G. Alonso, M.J. Franklin, W. Hong and J. Wisom, A pipelined framework for online cleaning of sensor data streams, in *Proceedings of the 22nd International Conference on Data Engineering (ICDE 06)*, 2006, p. 140.

[9] M. Balazinska et al., Data management in the worldwide sensor web, *IEEE Pervasive Computing* **6**(2) (April 2007), 30–40.

[10] K.G. Jeffery, The Internet of Things: the death of a traditional database? *IETE Technical Review* **26**(5) (2009), 313–319.

[11] J. Cooper and A. James, Challenges for database management in the Internet of Things, *IETE Technical Review* **26**(5) (2009), 320–329.

[12] Y. Diao, D. Ganesan, G. Mathur and P. Shenoy, Rethinking data management for storage-centric sensor networks, in *Proceedings of Third Biennial Conference on Innovative Data Systems Research (CIDR)*, January 2007, pp. 410–419.

[13] Z. Yu, B. Mo et al., Achieving optimal data storage position in wireless sensor network, *Computer Communications* **33**(1) (Jan. 2010), 92–102.

[14] M. Palmer, Seven Principles of Effective RFID Data Management, *Progress Software-Real Time Division*, 2004.

[15] F. Wang and P. Liu, Temporal management of RFID data, in *Proceedings of the 31st international conference on Very large data bases (VLDB)*, 2005, pp. 1128–1139.

[16] M. Navarro, D. Bhatnagar and Y. Liang, An integrated network and data management system for heterogeneous WSNs, in *Proceedings of 8th IEEE International Conference on Mobile Ad-Hoc and Sensor Systems*, 2011, pp. 819–824.

[17] T. Fan and Y. Chen, A scheme of data management in the Internet of Things, in *Proceedings of the 2nd IEEE International Conference on Network Infrastructure and Digital Content*, 2010, pp. 110–114.

[18] C. Fan et al., A scalable Internet of Things lean data provision architecture based on ontology, in *Proceedings of IEEE GCC Conference and Exhibition*, 2011, pp. 553–556.

[19] I. Buchan, J. Winn and C. Bishop, A unified modeling approach to data-intensive healthcare, *The fourth paradigm: data-intensive scientific discovery*, 2009, pp. 91–97

[20] M. Adnane, Z. Jiang, S. Choi and H. Jang, Detecting specific health-related events using an integrated sensor system for vital sign monitoring, *Sensors* **9**(9) (2009), 6897–6912.

[21] B. Logan, J. Healey, M. Philipose, E.M. Tapia and S. Intille, A long-term evaluation of sensing modalities for activity recognition, in *Proceedings of the 9th international conference on Ubiquitous computing*, 2007, pp. 483–500.

[22] T.L. Kasteren, G. Englebienne and B. Kröse, An activity monitoring system for elderly care using generative and discriminative models, *Personal and Ubiquitous Computing* **14**(6) (September 2010), 489–498.

[23] M. Philipose et al., Inferring activities from interactions with objects, *IEEE Pervasive Computing* **3**(4) (2004), 50–57.

[24] N.C. Krishnan and S. Panchanathan, Analysis of low resolution accelerometer data for continuous human activity recognition, in *Proceeding of IEEE International Conference on Acoustics, Speech and Signal Processing*, April 2008, pp. 3337–3340.

[25] U. Maurer, A. Smailagic, D.P. Siewiorek and M. Deisher, Activity recognition and monitoring using multiple sensors on different body positions, in *Proceedings of the International Workshop on Wearable and Implantable Body Sensor Networks*, 2006, pp. 113–116.

[26]   A. Helal et al., The Gator Tech smart house: A programmable pervasive space, *IEEE Computer* **38**(3) (March 2005), 50–60.

[27]   T. van Kasteren and B. Kröse, Bayesian activity recognition in residence for elders, in *Proceedings of the International Conference on Intelligent Environments*, 2008.

[28]   J. Lester, T. Choudhury, N. Kern, G. Borriello and B. Hannaford, A hybrid discriminative/generative approach for modeling human activities, in *Proceedings of the 19th international joint conference on Artificial intelligence*, 2005, pp. 766–772.

[29]   P. Rashidi, D.J. Cook, L.B. Holder and M. Schmitter-Edgecombe, Discovering activities to recognize and track in a smart environment, *IEEE Transactions on Knowledge and Data Engineering* **23**(4) (April 2011), 527–539.

[30]   D.H. Hu and Q. Yang, Transfer Learning for Activity Recognition via Sensor Mapping, in *Proceedings of the Twenty-Second International Joint Conference on Artificial Intelligence (IJCAI 2011)*, 2011, pp. 1962–1967.

[31]   D.H. Hu, X.X. Zhang, J. Xin, V.W. Zheng and Q. Yang, Abnormal Activity Recognition based on HDP-HMM Models, in *Proceedings of the Twenty-First International Joint Conference on Artificial Intelligence (IJCAI 2009)*, 2009, pp. 1715–1720.

[32]   J. Yin, D.H. Hu and Q. Yang, Spatio-temporal Event Detection Using Dynamic Conditional Random Fields, in *Proceedings of the Twenty-First International Joint Conference on Artificial Intelligence (IJCAI 2009)*, 1321–1326.

[33]   L. Chen, C.D. Nugent and H. Wang, A Knowledge-Driven Approach to Activity Recognition in Smart Homes, *IEEE Transactions on Knowledge and Data Engineering* **24**(6) (2012), 961–974.

[34]   D.J. Cook, N.C. Krishnan and P. Rashidi, Activity Discovery and Activity Recognition: A New Partnership, *IEEE Transactions on Systems, Man and Cybernetics, Part B (TSMCB)*, 2012, pp. 1–9.

[35]   P. Rashidi and D.J. Cook, A Method for Mining and Monitoring Human Activity Patterns for Home-based Health Monitoring Systems, *ACM Transactions on Intelligent Systems and Technology (TIST)*, *Special Issue on Intelligent Systems for Health Informatics*, 2012.

[36]   T. Gu, L. Wang, X. Tao and J. Lu, A Pattern Mining Approach to Sensor-based Human Activity Recognition, *IEEE Transactions on Knowledge and Data Engineering* **23**(9) (September 2011), 1359–1372.

[37]   T. Gu, L. Wang, X. Tao and J. Lu, RecognizingMulti-user Activities usingwearable sensors in a smart home, *IEEE Transactions on Mobile Computing (TMC)* **10**(11) (March 2011), 1618–1631.

[38]   T. Heath and C. Bizer, Linked Data: Evolving the Web into a Global Data Space, *Synthesis Lectures on the Semantic Web: Theory and Technology* **1**(1) (February 2011), 1–136.

[39]   H. Zhuge and Y. Sun, Schema Theory for Semantic Link Network, *Future Generation Computer Systems* **26**(3) (March 2010), 408–420.

[40]   J. Zhang and Y. Sun, An analogy reasoning model for Semantic Link Network, *JDCTA: International Journal of Digital Content Technology and its Applications* **4**(7) (2010), 128–139.

[41]   P. Barnaghi, W. Wang, C. Henson and K. Taylor, Semantics for the Internet of Things: early progress and back to the future, *International Journal on Semantic Web and Information Systems (IJSWIS)* **8**(1) (2012), 1–21.

[42]   E. Zhou, N. Zhong and Y. Li, Extracting news blog hot topics based on the W2T Methodology, *World Wide Web*, 2013, pp. 1–28.

[43]   R. Shaw, R. Troncy and L. Hardman, Lode: linking open descriptions of events, *The Semantic Web*, 2009, pp. 153–167.

[44]   Y. Raimond and S. Abdallah, The event ontology, 2007. [Online]. Available: http://purl.org/NET/c4dm/event.owl.

[45]   D. Guinard, V. Trifa, T. Pham and O. Liechti, Towards PhysicalMashups in the Web of Things, *in Proceedings of the 6th International Conference on Networked Sensing Systems (INSS'09)*, 2009, pp. 1–4.

**Yunchuan Sun** received his Ph.D. degree from the Institute of Computing Technology, Chinese Academy of Science, Beijing, China. He now works at the School of Economic and Business Administration, Beijing Normal University, Beijing, China. His current research interests include the theory and application of semantic link network and Internet of Things.

**Hongli Yan** received his B.S. degree in 2012 from Beijing Normal University. He is currently a postgraduate student in the College of Information Science and Technology. His current research interests include the Internet of Things, data mining and computational intelligence.

**Cheng Lu** received her B.S. degree in 2012 from Beijing Normal University. She is now a postgraduate student in the College of Information Science and Technology. Her current research interests include the Internet of Things, Mobile Social Networks and computational intelligence.

**Rong-Fang Bie** received her Ph.D. degree in 1996 from Beijing Normal University, where she is now a professor. She visited the Computer Laboratory at the University of Cambridge in 2003. Her current research interests include knowledge representation and acquisition for the Internet of Things, computational intelligence and model theory.

**Zhangbing Zhou** is an associate professor at the School of Information Engineering, China University of Geosciences (Beijing), and serves as an adjunct associate professor at the computer science department, Institut Mines TELECOM/TELECOM SudParis, France. He received his Ph.D. from the Digital Enterprise Research Institute (DERI), National University of Ireland, Galway (NUIG). His research interests include process-aware information system, service-oriented computing, spatial and temporal database, and sensor network middleware. Email: zhangbing.zhou@gmail.com.

# 7

# Diffie-Hellman key based authentication in proxy mobile IPv6

HyunGon Kim[a] and Jong-Hyouk Lee[b,*]

[a]*Department of Information Security, Mokpo National University, Korea*
[b]*IMARA Team, INRIA, Paris, France*

**Abstract.** Wireless communication service providers have been showing strong interest in Proxy Mobile IPv6 for providing network-based IP mobility management. This could be a prominent way to support IP mobility to mobile nodes, because Proxy Mobile IPv6 requires minimal functionalities on the mobile node. While several extensions for Proxy Mobile IPv6 are being developed in the Internet Engineering Task Force, there has been little attentions paid to developing efficient authentication mechanisms. An authentication scheme for a mobility protocol must protect signaling messages against various security threats, e.g., session stealing attack, intercept attack by redirection, replay attack, and key exposure, while minimizing authentication latency. In this paper, we propose a Diffie-Hellman key based authentication scheme that utilizes the low layer signaling to exchange Diffie-Hellman variables and allows mobility service provisioning entities to exchange mobile node's profile and ongoing sessions securely. By utilizing the low layer signaling and context transfer between relevant nodes, the proposed authentication scheme minimizes authentication latency when the mobile node moves across different networks. In addition, thanks to the use of the Diffie-Hellman key agreement, pre-established security associations between mobility service provisioning entities are not required in the proposed authentication scheme so that network scalability in an operationally efficient manner is ensured. To ascertain its feasibility, security analysis and performance analysis are presented.

Keywords: PMIPv6, NETLMM, authentication, security

## 1. Introduction

While conventional mobility solutions have been developed based on host-based mobility management, the concept of network-based mobility management has been introduced in the Internet Engineering Task Force (IETF). Because conventional mobility solutions such as Mobile IPv6 (MIPv6) [8] and its extensions force a mobile node (MN) to have heavy functionalities for supporting its own mobility service, wireless communication service providers turn their gaze on network-based mobility management [3, 14,15]. Proxy Mobile IPv6 (PMIPv6) [21] is a recently developed mobility protocol from the concept of network-based mobility management wherein mobility service for an MN is provided by mobility service provisioning entities. The mobility service provisioning entities in the PMIPv6 domain manage all mobility signaling and data structures for the MN. Accordingly, an ordinary MN, which does not implement the mobility stack required in the conventional mobility solutions, achieves its mobility service in the given PMIPv6 domain [14,21].

In the IETF, extensions for PMIPv6 are being actively developed. Especially, several fast handover mechanisms proposed to minimize handover latency are introduced. Then, Fast Handovers for Proxy

---

*Corresponding author: Jong-Hyouk Lee, IMARA Team, Bt. 07, INRIA Paris – Rocquencourt, Domaine de Voluceau Rocquencourt, B.P. 105, 78153, Le Chesnay Cedex, France. Tel.: +33 1 39 63 59 30; E-mail: jong-hyouk.lee@inria.fr.

Mobile IPv6 (FPMIPv6) [11] proposed by Yokota et al. has been selected and being standardized in the IETF. Even if FPMIPv6 has been well introduced how to reduce handover latency and packet loss while an MN moves different networks in the given PMIPv6 domain, it does not consider security issues. In other words, the MN must undergo its authentication procedure to have network access authorization when it attaches to a new network [2,13], but FPMIPv6 does not supply to reduce authentication latency occurred when the MN changes its access network.

In order to provide authenticated handover service for authorized MNs, an authentication, authorization, and accounting (AAA) architecture is a major security architecture [4,6,10] that has been widely being used in the networks of wireless communication service providers. Accordingly, it is naturally expected that PMIPv6 will be deployed in many networks with the AAA architecture [13]. However, the base specification of PMIPv6 has been developed with a limited understanding of secure authentication and actual deployment scenarios. For instance, 1) the impact of handover authentication is not addressed even if it contributes as an important performance metric, 2) the impact of the chosen integrity, confidentiality, and authentication methods is not addressed. We therefore need a secure handover scheme considering efficient secure authentication elements and deployment scenarios in order to deploy PMIPv6 mobility service within the AAA architecture successfully.

In this paper, we introduce a Diffie-Hellman (DH) key based authentication scheme that utilizes the low layer signaling to exchange DH variables and allows mobility service provisioning entities to exchange mobile node's profile and ongoing sessions securely. More precisely, the introduced DH key based authentication scheme has the following distinctive features compared to PMIPv6.

- DH key exchange operation is adopted to reduce the computation overhead.
- Relevant mobility service provisioning entities are supported to perform the context transfer and data packet forwarding.
- Pre-established security associations between mobility service provisioning entities are not required.

By utilizing the distinctive features, the DH key based authentication scheme achieves low handover latency while providing secure handover service for MNs in PMIPv6. The current specifications of PMIPv6 and FMIPv6 only provide the protocol operations without secure authentication concerns. Accordingly, the proposed DH key based authentication scheme would be a good direction for secure authentication for PMIPv6.

The remainder of this paper is organized as follows: Section 2 describes the specification of PMIPv6 and FPMIPv6 with the operation scenario within the AAA architecture. Then, in Section 3, the proposed DH key based authentication scheme is presented with the protocol operation and the security analysis. In Section 4, the results of performance evaluation are presented compared to existing authentication schemes. The conclusions of this paper are presented in Section 5.

## 2. Related work

In this section, we present the basic operation of PMIPv6 defined in [21]. Then, its extension, FPMIPv6 [11], is also described. PMIPv6 only defines the basic handover operation and data structure for network-based mobility management, whereas FPMIPv6 applies the concept of fast handover into PMIPv6 to improve handover performance.

## 2.1. Proxy Mobile IPv6

PMIPv6 is a network-based mobility management protocol reusing MIPv6 entities and concepts as much as possible. The core functional entities, i.e., mobility service provisioning entities, are the mobile access gateway (MAG) and the local mobility anchor (LMA). The MAG is usually located at the access router (AR) as software functionalities. The MAG detects the movement of an MN and then sends a proxy binding update (PBU) message to the LMA in order to register the location information of the MN. It means that mobility service for the MN is supported by the mobility service provisioning entities such as the MAG and the LMA. When the LMA receives the PBU message including essential information for the MN, the LMA recognizes that the MN has been attached to the access network managed by the MAG. As a response, the LMA sends back a proxy binding acknowledgement (PBAck) message including the home network prefix (HNP) for the MN. Then, the MAG sends a router advertisement (RA) message including the HNP to the MN. The LMA and the MAG establishes a bi-directional tunnel for the MN. Because that the LMA is responsible for maintaining the MN's reachability state and is the topological anchor point for the MN's HNP, all data packets sent from and to the MN are smoothly delivered through the established bi-directional tunnel for the MN.

As the MN receives the RA message sent from its MAG, the MN configures its address based on the HNP included in the RA message. Compared to MIPv6, this configured address is not a care-of address (CoA), which is changed when an MN changes its point of attachment in MIPv6, but it is treated as a home address (HoA). In other words, the MN continuously obtains and uses the same address called as a Proxy-HoA in the given PMIPv6 domain. This is because that the LMA continuously provides the same HNP for the MN.

In PMIPv6, each MN must be identified by its identifier, MN-ID. The MN-ID is used to obtain the MN's profile describing the allowed LMA's address (LMAA), i.e., MN-LMAA, assigned HNP, i.e., MN-HNP, permitted address configuration mode, roaming policy, and other parameters. Note that the MN's profile is an abstract term for referring to a set of configuration parameters configured for the given MN. The mobility service provisioning entities in the PMIPv6 domain are thus required to access these parameters in order to provide the mobility service for the MN. This information (profile) is typically stored in a policy store at an AAA server. Accordingly, upon completing the authentication procedure for the MN, this profile is retrieved and used to execute network-based mobility management for the MN.

Here, we give an actual example of the handover procedure of PMIPv6. Suppose that the MN has attached with the MAG$_1$ and changes its point of attachment to the MAG$_2$ in the same PMIPv6 domain managed by the home LMA (LMAh).

As illustrated in Fig. 1, there are several message exchanging between nodes. The detailed descriptions for the message exchanging are as follows:

1. De-registration Proxy Binding Update (De-Reg. PBU) message: By utilizing the L2 trigger, the MAG$_1$ detects that the MN will change its point of attachment. Then, the MAG$_1$ sends the De-registration Proxy Binding Update (De-Reg. PBU) message to the LMAh in order to inform the detachment of the MN on the access network. The binding and routing state for the MN is removed at the binding update list at the MAG.

2. De-registration Proxy Binding Update Acknowledgement (De-Reg. PBAck) message: Upon receiving the De-Reg. PBU message indicating that the MN has been detached from that access network, the LMAh checks its corresponding mobility session for the MN and accepts the De-Reg. PBU message if it is valid. Then, the LMAh waits for a pre-defined time to allow the MAG on the

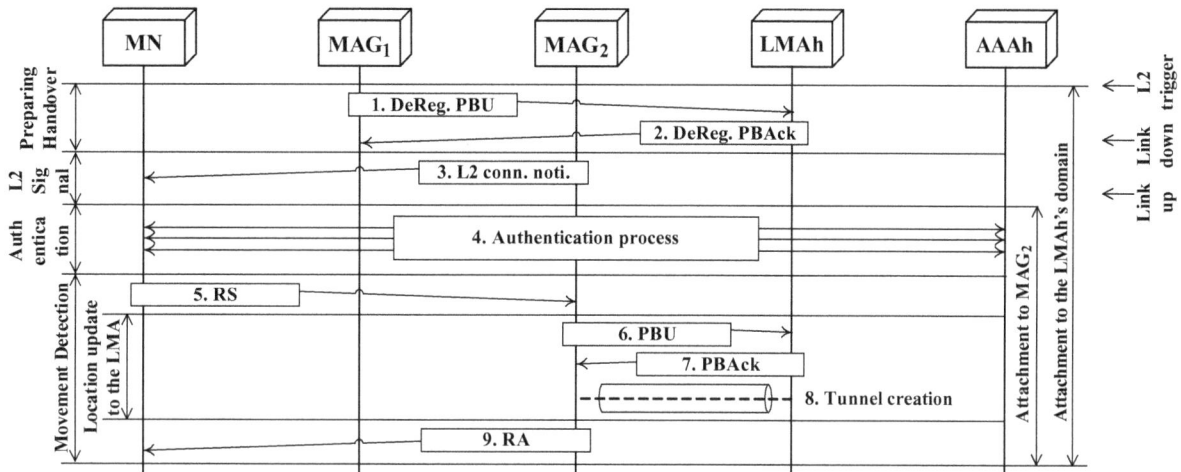

Fig. 1. The handover procedure of PMIPv6.

new access network to update the binding of the MN. That is, the LMAh waits for receiving a PBU message for the MN for a certain amount of time.

3. L2 Connection Notification: As the MN approaches the $MAG_2$, it receives the L2 connection notification. And then, the MN's wireless interface will be attached to the $MAG_2$.

4. Authentication Process: An authentication process for authorizing the MN in the new access network must be placed before the MN makes actual communication sessions. In Fig. 1, the exact authentication process it not presented, but a strong authentication mechanism must be applied when PMIPv6 is deployed in real network environments. For instance, the EAP-based authentication framework or public-key based framework with the AAA architecture can be used in here. Note that the AAA home server (AAAh) appears in this example.

5. Router Solicitation (RS) message: As soon as successful authentication of the MN, the MN sends the RS message in order to explicitly inform its attachment to the $MAG_2$ and to receive the RA message quickly.

6. Proxy Binding Update (PBU) message sent from the $MAG_2$: As receiving the RS message, the $MAG_2$ recognizes the presence of the MN and then sends the PBU message to the LMAh. In PMIPv6, the movement detection of the MN can be achieved in several ways. For instance, 1) by utilizing the L2/L3 signaling, MAGs can detect the movement of the MN and 2) by utilizing Authentication, Authorization and Accounting (AAA) signaling, MAGs can detect the movement of the MN as well as it obtains extra information of the MN. In [9,16], AAA operations and examples for PMIPv6 have been presented.

7. Proxy Binding Acknowledgement (PBAck) message: The LMAh on receiving the PBU message sent from the $MAG_2$ recognizes that the MN has been attached to the $MAG_2$. The LMAh sends the PBAck message including the same HNP that has been assigned to the MN at the previous access network of the MN.

8. Bidirectional Tunnel: As receiving the PBAck message indicating the success of the binding for the MN, a bidirectional tunnel between the LMAh and the $MAG_2$ is established for data packets forwarding for the MN.

9. Router Advertisement (RA) message: Because the LMAh assigns the same HNP for the MN, the MN is ensured to receive the same HNP compared to the previous one. The MN continuously finds

the same HNP in the RA message. The MN therefore configures and uses the same Proxy-HoA in the LMAh's domain.

Throughout the message exchanging, the MN is allowed to change its point of attachment without its actual involvement in mobility signaling actions, e.g., sending a message to register its new location. Compared to the previously developed mobility protocols such as MIPv6, PMIPv6 has a simple but devoted mobility support for the MN.

The limitations that the base PMIPv6 specification has are 1) route optimization (RO) support, 2) multihomed MN support, and 3) handover optimization. The recently chartered Network-Based Mobility Extensions (NETEXT) working group [19] are currently working on the issues of RO support and multihomed MN support in PMIPv6. Then, the issues of handover optimization is currently treated in Mobility for IP: Performance, Signaling and Handoff Optimization (MIPSHOP) working group [18]. FPMIPv6 [11] proposed by Yokota et al. has been being developed as a working group item in the MIPSHOP working group.

## 2.2. Fast proxy mobile IPv6

FPMIPv6 has been introduced in order to minimize the handover latency incurred while an MN performs its handover. The base specification of PMIPv6 does not cover this issues. As presented in [11], fast handover mechanisms introduced in MIPv6 called FMIPv6 [20] cannot be directly applied into PMIPv6 because that the MN cannot launch any mobility signaling to indicate its movement to ARs.

Similar to FMIPv6, FPMIPv6 operates in either the predictive mode and the reactive mode depending on the network circumstances. More precisely, if the MN does not have enough time to prepare its handover at the currently attached network, the reactive mode is activated, whereas the predictive mode is launched when the MN moves to the new access network after the completion of the context transfer between the relevant ARs. Obviously, the predictive mode can significantly reduce the handover latency compared to the reactive mode. The reactive mode hardly reduces the handover latency. Therefore we here focus on the predictive mode of FPMIPv6 because it actually achieves the goal of fast handover.

Figure 2 illustrates the message exchanging between nodes for the handover procedure of Predictive FPMIPv6. In Fig. 2, the MN prepares its handover at the currently attached network managed by the $MAG_1$ to the new network managed by the $MAG_2$. The detailed descriptions for the message exchanging are as follows:

1. L2 report: By utilizing the L2 trigger, the MN detects information of neighbor networks. Then, it reports information of the next network the MN will attach to with. This message is a access technology specific, but at least the MN-ID and network identification (NET-ID) must be provided to the $MAG_1$, where the MN is currently attached with.

2. Handover Initiate (HI) message: As receiving the L2 report sent from the MN, the $MAG_1$ recognizes that the MN will move to the specific network indicated by the NET-ID, i.e., the network managed by the $MAG_2$. Then, it informs the movement of the MN to the $MAG_2$ by sending the handover initiate (HI) message including the MN-ID,MN-HNP, MN-LMAA, MN's interface identification (MN-IID), etc.

3. Handover Acknowledge (HAck) message: The $MAG_2$ replies with the handover acknowledge (HAck) message indicating the success or failure for preparing of the MN's handover.

4. De-registration Proxy Binding Update (De-Reg. PBU) message: The $MAG_1$ sends the De-Reg. PBU message to the LMAh in order to inform the detachment of the MN on the access network. Depending on the actual implementation, the HI and De-Reg. PBU messages can be simultaneously sent, but in Fig. 2, it has been presented in stepwise.

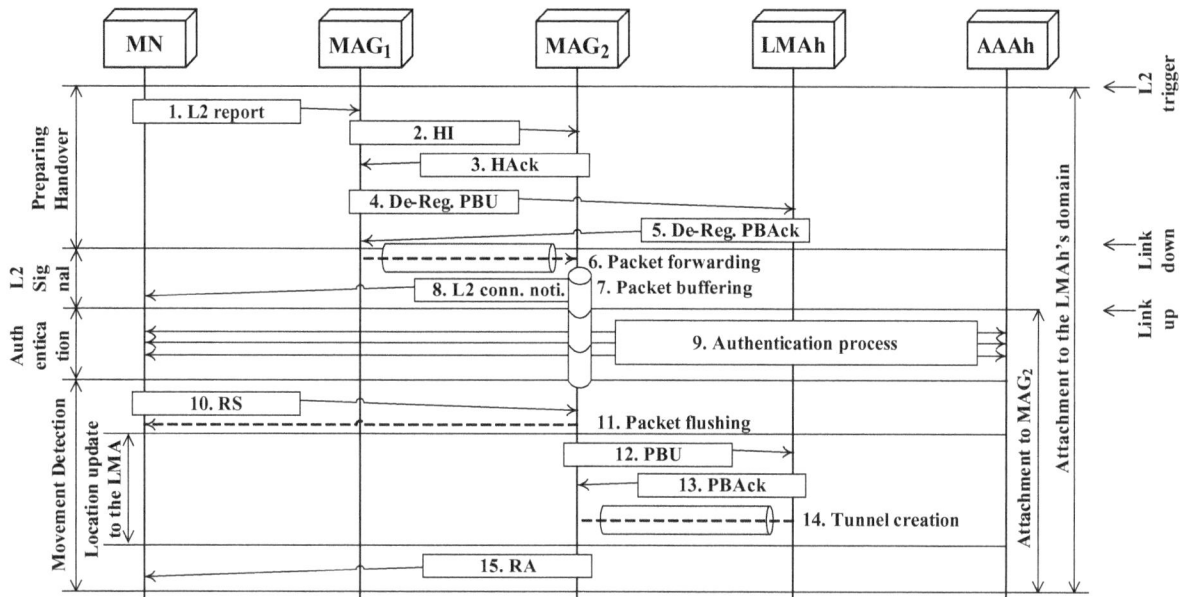

Fig. 2. The handover procedure of Predictive FPMIPv6.

5. De-registration Proxy Binding Update Acknowledgement (De-Reg. PBAck) message: Upon receiving the De-Reg. PBU message indicating that the MN has been detached from that access network, the LMAh checks its corresponding mobility session for the MN and accepts the De-Reg. PBU message if it is valid. Then, the LMAh waits for a pre-defined time to allow the MAG on the new access network to update the binding of the MN.

6. Packet Forwarding: The $MAG_1$ starts to forward data packets destined for the MN.

7. Packet Buffering: The data packets forwarded from the $MAG_1$ are being buffered at the $MAG_2$. The actual implementation and operation of packet forwarding and buffering can be different depends on the implementation details, but the goal of packet forwarding and buffering is to prevent packet loss.

8. L2 Connection Notification: As the MN approaches the network of $MAG_2$, it receives the L2 connection notification from the $MAG_2$. Then, the MN's wireless link is attached to the $MAG_2$.

9. Authentication Process: Similar to PMIPv6, the authentication process introducing unacceptable long latency is preformed if it is not optimized.

10. Router Solicitation (RS) message: As soon as successful authentication of the MN, the MN sends the RS message in order to explicitly inform its attachment to the $MAG_2$ and to receive the RA message quickly.

11. Packet Flushing: The $MAG_2$ immediately sends the data packets to the MN. However, this packet flushing can begin as the authentication process is done. This is, data packets can be sent before the $MAG_2$ receives the RS message if the $MAG_2$ explicitly knows the attachment of the MN at its access network throughout other information, i.e., the authentication success message for the MN sent from the AAAh server when it acts as an AAA client.

In Fig. 2, other message exchanging which has been not described here is similar with that of PMIPv6. Predictive FPMIPv6 obviously reduces the handover latency by allowing the MN prepares its handover before the MN performs its actual handover to the new access network. The HI and HAck message presented in Fig. 2 are used to exchange the context transfer between the MAGs. Then, as the MN attaches to the new network managed by one of the MAGs, i.e., $MAG_2$, the buffered data packets for the MN are immediately sent to the MN.

Even if FPMIPv6 improves handover performance of PMIPv6, it cannot address the handover authentication latency occurred during the MN undergoes its authentication process. For instance, the required times for several message exchanging between the MN and the AAAh, and executing cryptography operation yield long latency. This long latency for handover authentication thus causes user-perceptible deterioration of handover performance even if FPMIPv6 is used. The objective of this paper is to reduce such long handover authentication latency.

## 3. Diffie-hellman key based authentication

In this section, we present the proposed DH key based authentication scheme. The followings are the design principles and assumptions of the DH key based authentication scheme.

- Minimizing the computation power consumption as well as the administrative cost imposed on the MN.
- Minimizing the number of keying material requests to the AAAh.
- Utilizing signaling messages defined in FPMIPv6 to improve handover performance.
- Utilizing L2 events to anticipate the handover of the MN.
- Utilizing L2 messages in order to carry DH variables.
- Protecting session keys against various attacks.
- Removing pre-established security associations between the MAGs.
- Removing additional signaling messages between the MAG and the LMA.

One of recent performance enhancement approaches is to use link-layer specific information. For instance, IEEE 802.21 (MIH) provides link-layer specific information to upper layers. Especially, some information provided by IEEE 802.21 such as available network list, link identification, link status, etc, can be used to facilitate the handover decision and detection of the MN [1,7]. In this paper, we assume that the MN and network entities are aware of MIH functionalities.

### 3.1. Protocol operation

Figure 3 depicts the message exchanging between nodes for the handover procedure of the proposed DH key based authentication scheme. For variant DH key exchange operations, MAGs choose a large prime number $n$, generate $g \in Z_n$ and $y \in Z_{n-1}$ at random, and compute $g^y \in Z_n$ in advance. The detailed descriptions are as follows:

1. L2 signal ($g_{old}^{y'}$, $g_{old}$, $n_{old}$) from the $MAG_1$: As a beacon signaling, the $MAG_1$ sends the L2 signal. This message contains $g_{old}^{y'}$, $g_{old}$, and $n_{old}$. In addition, this message includes the available network list, link identification, link status, etc.

2. L2 signal ($g_{new}^{y}$, $g_{new}$, $n_{new}$) from the $MAG_2$: Similarly, the $MAG_2$ sends the L2 signal containing $g_{new}^{y}$, $g_{new}$, and $n_{new}$.

3. L2 report ($g_{new}^{y}$, $g_{old}^{y'}$, $g_{new}$, $n_{new}$): The MN sends the L2 report including $g_{new}^{y}$, $g_{old}^{y'}$, $g_{new}$, and $n_{new}$. In addition, this message includes at least the MN-ID and NET-ID. As receiving this message, the $MAG_1$ recognizes that the MN is going to attach with the network managed

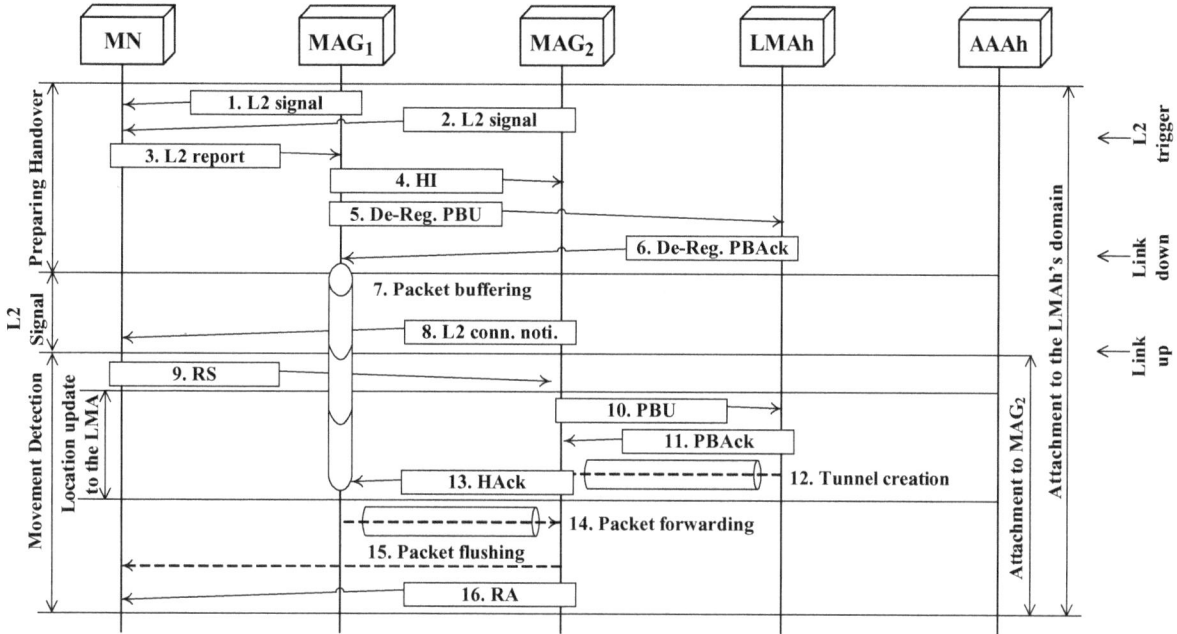

Fig. 3. The handover procedure of the proposed scheme.

by the $MAG_2$. Now, the $MAG_1$ computes $x = <g_{new}^y, g_{old}^{y'}>S_{MAG-LMA}(\text{mod}\,n_{new-1})$ and creates $g_{new}^x \in Z_{n_{new}}$. Also, the $MAG_1$ creates a session key $K_{MAG_1-MAG_2} = (g_{new}^y)^x = g_{new}^{xy} \in Z_{n_{new}}$. Finally, the $MAG_1$ encrypts $C = [S_{MN-MAG}, S_{MAG-LMA}]K_{MAG_1-MAG_2}$ and $C' = [M_\rho, K_\chi]K_{MAG_1-MAG_2}$, where $M_\rho$ is the profile's MN and $K_\chi$ is the ongoing mobility session key for the MN.

4. Handover Initiate (HI) message $(C, C', g_{new}^y, g_{old}^{y'})$: As receiving the HI message including $C, C'$, $g_{new}^y$, and $g_{old}^{y'}$, the $MAG_2$ validates $g_{new}^y$ and stores $C$. Then, it retrieves $g_{new}^y, g_{new}$, and $n_{new}$.

5. De-registration Proxy Binding Update (De-Reg. PBU) message: As a default operation defined in [21], this message is sent to the LMA in order to inform the detachment of the MN.

6. De-registration Proxy Binding Update Acknowledgement (De-Reg. PBAck) message: As a response to the De-Reg. PBU message, it is sent from the LMAh to the $MAG_1$.

7. Packet Buffering: The $MAG_1$ starts to buffer data packets destined for the MN. This packet buffering is continued until the $MAG_1$ receives the HAck message.

8. L2 Connection Notification: As the MN approaches the network of $MAG_2$, it receives the L2 connection notification.

9. Router Solicitation (RS) message: The MN sends the RS message in order to explicitly inform its attachment to the $MAG_2$ and to receive the RA message quickly.

10. Proxy Binding Update (PBU) message $(g_{new}^y, g_{old}^{y'}, g_{new}, n_{new})$: As receiving the RS message, the $MAG_2$ knows the attachment of the MN. Then, it sends the PBU message including $g_{new}^y, g_{old}^{y'}$, $g_{new}, n_{new}$ to the LMAh. In addition, it computes $x = <g_{new}^y, g_{old}^{y'}>S_{MAG-LMA}(\text{mod}\,n_{new-1})$ and creates $g_{new}^x \in Z_{n_{new}}$.

11. Proxy Binding Acknowledgement (PBAck) message $(g_{new}^x)$: Once the LMAh successfully processes the PBU message, it replies the PBAck message including $g_{new}^x$. The $MAG_2$ decrypts $C$ and then obtains $S_{MN-MAG}$ and $S_{MAG-LMA}$. In addition, it decrypts $C'$ and then obtains

$M_\rho$ and $K_\chi$. Accordingly, the $MAG_2$ now obtains all materials for serving the MN at its access network.

12. Bidirectional Tunnel: The bidirectional tunnel is established for the MN.
13. Handover Acknowledge (HAck) message: The $MAG_2$ sends the HAck message indicating its successful handover authentication.
14. Packet Forwarding: The $MAG_1$ now starts to forward data packets destined for the MN if the HAck message indicates the success of the handover authentication.
15. Packet Flushing: The $MAG_2$ immediately forwards the data packets to the MN.
16. Router Advertisement (RA) message: The RA message including the same HNP compared to the previous one is sent to the MN.

In the proposed scheme, the previously assigned session keys $S_{MN-MAG}$ and $S_{MAG-LMA}$ are reused to reduce the key generation time and the key delivery time. This feature the proposed scheme has avoids the contact with the AAAh for authentication of the MN every time the MN changes its point of attachment. To ensure the confidentiality and integrity of these session keys, they are encrypted and decrypted under a short-term secret key $K_{MAG_1-MAG_2}$. In order to provide mobility service for the newly attached MN, the network must obtain related information for the MN. In the proposed scheme, $M_\rho$ and $K_\chi$ are also forwarded from the previous network to the new network. To minimize the MN's computing power consumption, the MAGs create $n$ and $g \in Z_n$. Since the MAGs use $n$ and $g$ for only a short period, at most 512 bits prime $n$ should be large enough. It results in reducing the computational overhead for $g^x$ and $g^y$.

## 3.2. Security analysis

In this section, we present security analysis results on the proposed DH key based authentication scheme.

The proposed scheme reuses the previously assigned session keys to achieve low handover latency. However, security weaknesses about this key reuse must be addressed. Accordingly, we point out a possible session-stealing attack on the proposed scheme. In order to re-use the session keys, however, they have to be taken over in a secure fashion between the relevant MAGs. Especially, $S_{MAG-LMA}$ is a random value of at least 64 bits and is not hashed. Unfortunately, if an attacker spoofs at the new network managed by the $MAG_2$, he can acquire the keys in the phase of key exchange between the $MAG_1$ and the $MAG_2$, and then the current session can be derived from that key. Because of this security weakness, the confidentiality of the session keys must be provided in the phase of key exchange. For a similar purpose, the solution proposed by Jacobs and Belgard [22] can be viewed as a further attempt to provide confidentiality of session keys based on public key cryptography. However, it is impractical because that each MAG must perform public key cryptography operations that suffer from a long delay during the handover authentication for the MN. The lifetime of the session keys is enough to avoid too-frequent AAA related transactions since each invocation of this process is likely to cause lengthy delays. Once the keys have been distributed by the AAAh, the $MAG_1$ obtains two session keys: $S_{MN-MAG}$ and $S_{MAG-LMA}$. If the MN attaches with the $MAG_2$, the PBU message is launched to update the location of the MN by the $MAG_2$. Since the $MAG_2$ has no session keys, re-authentication is required and new session keys should be assigned by the AAAh, which leads to long signaling delay. The proposed scheme thus uses existing session keys when there is enough key lifetime remaining in the existing binding update. This can eliminate the time required for re-authentication by the AAAh.

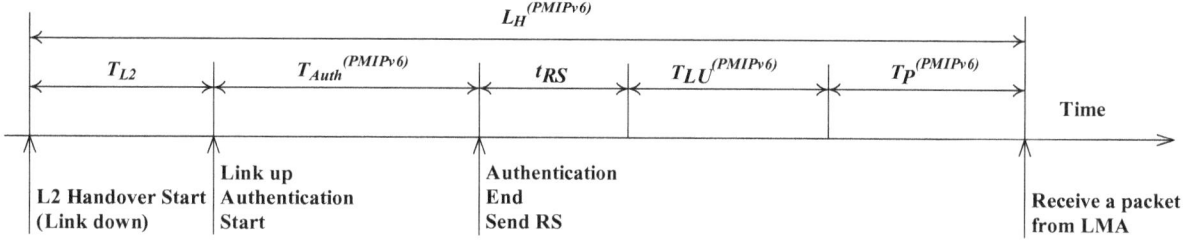

Fig. 4. The timing diagram for PMIPv6 handover.

Let us consider some scenarios considering possible session-stealing attacks from a session-stealing attacker's point of view: First, suppose an attacker intercepts the L2 report including $g_{new}^y, g_{old}^{y'}, g_{new}$ and $n_{new}$. The attacker can obtain the encrypted message $C$ and $g^y$, but cannot decrypt $C$ since he does not have $K_{MAG_1-MAG_2}$. Furthermore, the attacker cannot compute $K_{MAG_1-MAG_2} = g^{xy}$ since he does not know $x$, even if he knows $g^y$. Second, suppose that an attacker intercepts the HI message including $C$, $C'$, $g_{new}^y$ and the PBAck message including $g_{new}^x$. Then, he only knows $C$, $g^x$ and $g^y$ so that he cannot compute $K_{MAG_1-MAG_2} = g^{xy}$ from $g^x$ and $g^y$ within the lifetime of the session keys since the DH problem is computationally infeasible. Therefore, we can assert that the proposed scheme provides confidentiality and integrity of ongoing session keys and enables MN's profile to be exchanged securely.

## 4. Performance analysis

In this section, we develop an analytical model to investigate the handover latency and the handover blocking probability. Then, we present the numerical results.

### 4.1. Handover latency

We define the handover latency as the time interval during which an MN cannot send or receive any packets while it performs its handover between different networks.

Figure 4 illustrates the timing diagram for PMIPv6 handover. PMIPv6 manages the movement of an MN in a localized manner, but the handover authentication for the MN must be performed for every time the MN changes its point of attachment in the given PMIPv6 domain.

Suppose $L_H^{(PMIPv6)}$ is the handover latency of PMIPv6. Then it is expressed as follows.

$$L_H^{(PMIPv6)} = T_{L2} + T_{Auth}^{(PMIPv6)} + t_{RS} + T_{LU}^{(PMIPv6)} + T_P^{(PMIPv6)}, \tag{1}$$

where $T_{L2}$ is the link-layer handover latency. This latency varies among different implementation chipsets. $T_{Auth}^{(PMIPv6)}$ is the handover authentication latency that can be estimated the transmission latency between the serving MAG for the MN and the AAAh. In the paper, the computation times for generating and verifying keys are assumed to be negligible. The transmission latency for delivering required keys from the AAAh is assumed to be a main factor for authentication latency. $t_{RS}$ is the required time for receiving the RS message sent from the MN so that it can be rewritten as $t_{MAG-MN}$, where $t_{MAG-MN}$ is the one-way transmission latency between the MAG and the MN. $T_{LU}^{(PMIPv6)}$ is the location update latency for the MN. It can be expressed as $2t_{MAG-LMA}$, where $t_{MAG-LMA}$ is the one-way transmission latency between the MAG and the LMA. $T_P^{(PMIPv6)}$ is the time required that

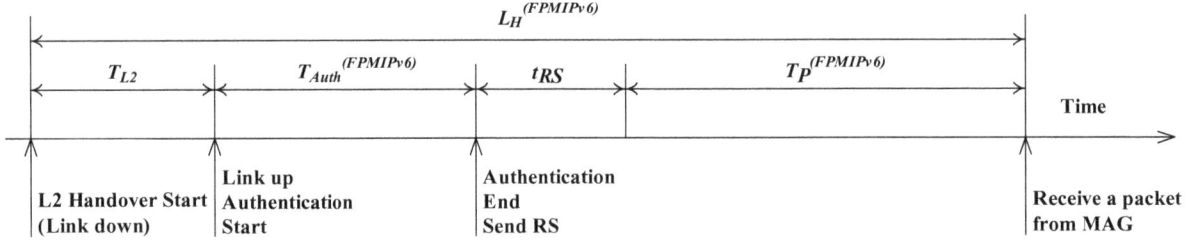

Fig. 5. The timing diagram for Predictive FPMIPv6 handover.

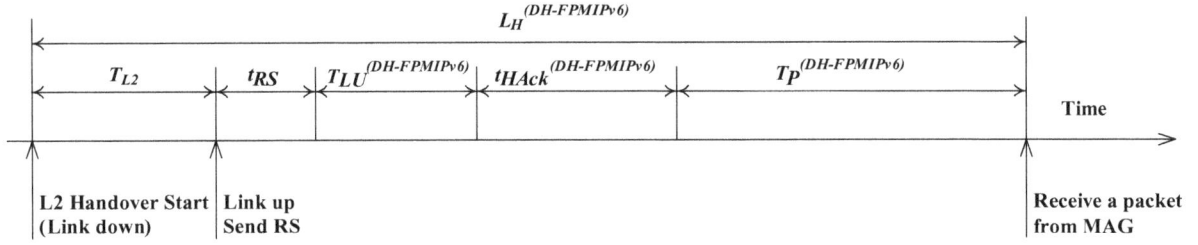

Fig. 6. The timing diagram for the proposed scheme's handover.

the first packet destined for the MN is arrived from the LMA to the MN. Then, it can be expressed as $t_{MAG-LMA} + t_{MAG-MN}$.

Figure 5 illustrates the timing diagram for Predictive FPMIPv6 handover. FPMIPv6 enables an MN prepares its handover before the MN performs its actual handover. Accordingly, the handover performance compared to that of PMIPv6 is improved. However, Predictive FPMIPv6 does not optimize the handover authentication. Accordingly, similar to PMIPv6, the handover authentication for the MN must be performed for every MN's handover.

Suppose $L_H^{(FPMIPv6)}$ is the handover latency of Predictive FPMIPv6. Then it is expressed as follows.

$$L_H^{(FPMIPv6)} = T_{L2} + T_{Auth}^{(FPMIPv6)} + t_{RS} + T_P^{(FPMIPv6)}, \qquad (2)$$

$T_{Auth}^{(FPMIPv6)}$ is the handover authentication latency of Predictive FPMIPv6 which is not different with that of PMIPv6 so that $T_{Auth}^{(FPMIPv6)}$ is the same with $T_{Auth}^{(PMIPv6)}$. $T_P^{(FPMIPv6)}$ is the time which the first data packet sent from the new network, i.e., MAG$_2$, arrives at the MN. The data packets destined for the MN have been buffered at the MAG$_2$ before the MN attaches with the MAG$_2$. As the MAG$_2$ recognizes the attachment of the MN by receiving the RS message, it immediately sends the buffered data packets to the MN. Accordingly, $T_P^{(FPMIPv6)}$ can be rewritten as $t_{MAG-MN}$.

Figure 6 illustrates the timing diagram for the proposed scheme's handover. In the proposed scheme, the authentication process can be done as the MAG$_2$ receives the PBAck message sent from the LMAh. In addition, the proposed one utilizes the buffering mechanism used in FPMIPv6 so that data packets being buffered at the MAG$_1$ in the previous network are forwarded as the MAG$_1$ receives the HAck message sent from the MAG$_2$.

Suppose $L_H^{(DH-FPMIPv6)}$ is the handover latency of the proposed scheme. Then it is expressed as follows.

$$L_H^{(DH-FPMIPv6)} = T_{L2} + t_{RS} + T_{LU}^{(DH-FPMIPv6)} + t_{HAck}^{(DH-FPMIPv6)} + T_P^{(DH-FPMIPv6)}, \qquad (3)$$

where $T_{LU}^{(DH-FPMIPv6)}$ is the location update latency that can be expressed as $2t_{MAG-LMA}$. $t_{HAck}^{(DH-FPMIPv6)}$ is the time which the HAck message sent from the $MAG_2$ arrives at the $MAG_1$. Then, it can be expressed as $t_{MAG-MAG}$, which is the one-way transmission latency between the MAGs. As receiving the HAck message, the $MAG_1$ immediately forwards data packets to the $MAG_2$ where the forwarded data packets for the MN are also sent to the MN. Accordingly, $T_P^{(DH-FPMIPv6)}$ is expressed as $t_{MAG-MAG} + t_{MAG-MN}$.

## 4.2. Handover blocking probability

The handover event will be failed due to several reasons. In this paper, we only consider the handover latency as a handover blocking factor. Then, the handover failure event can be expressed as the event which an MN cannot complete its handover when the network residence time is less than the handover latency.

Suppose $E[L_H^{(\cdot)}]$ is the mean value of $L_H^{(\cdot)}$, where $(\cdot)$ is a protocol indicator. For the sake of simplicity, we assume that $T_H^{(\cdot)}$ is exponentially distributed with its cumulative function $F_T(t)$. Then, the handover blocking probability $\rho_b^{(\cdot)}$ can be expressed as follows [12,23].

$$\rho_b^{(\cdot)} = Pr(L_H^{(\cdot)} > T_R) = \int_0^\infty (1 - F_T^{(\cdot)}(u)) f_R(u) du = \frac{\mu_c E[L_H^{(\cdot)}]}{1 + \mu_c E[L_H^{(\cdot)}]}, \tag{4}$$

where $T_R$ is the network residence time and its probability density function is $f_R(t)$. $\mu_c$ is the border crossing rate for the MN. Assuming that the AR's coverage area is circular, then $\mu_c$ is calculated as follows [5].

$$\mu_c = \frac{2\nu}{\pi R}, \tag{5}$$

where $\nu$ is the average velocity of the MN and $R$ is the radius of the AR's coverage area.

## 4.3. Numerical results

For the numerical analysis, we use the following system parameters obtained from the previous works [13,17]: $T_{L2} = 45.35$ ms, $t_{MAG-MN} = 12$ ms, $t_{MAG-MAG} = 15$ ms, and $t_{MAG-LMA} = 20$ ms.

In Fig. 7, we investigate the variation of the handover latency. As functions for the variation of the handover latency, we use $T_{Auth}$ and the number of handovers $n$. Figure 7(a) presents the variation of the handover latency as a function of $T_{Auth}$. From the presented results in Fig. 7(a), we can find that the proposed authentication scheme is not affected by $T_{Auth}$, but other schemes are affected. This is because that the proposed authentication scheme reuses the previously assigned session keys for an MN when the MN performs its handover from the previous network to the new network. The session keys used in the previous network are securely transferred to the new network. In other words, the proposed authentication scheme does not require to contact with the AAAh in order to authenticate the MN. Moreover, as we can see in Fig. 7(a), when $T_{Auth}$ is enough small value, Predictive FPMIPv6 outperforms other schemes, but as $T_{Auth}$ increases, the proposed authentication scheme shows the best performance compared to others. Next, we vary $n$ from 0 to 10 and fix $T_{Auth}$ as 80 ms. Then, we see

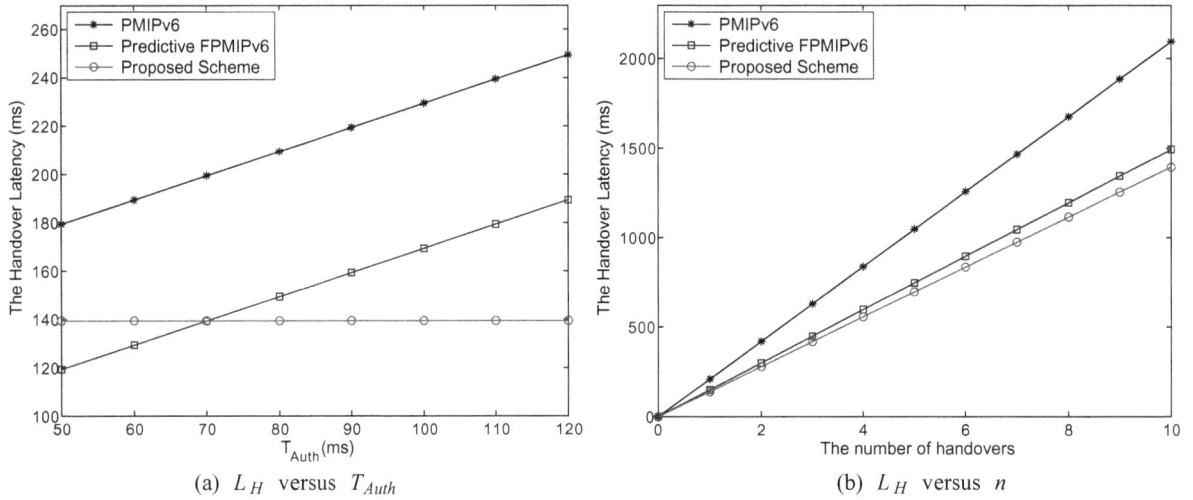

(a) $L_H$ versus $T_{Auth}$

(b) $L_H$ versus $n$

Fig. 7. The variation of the handover latency.

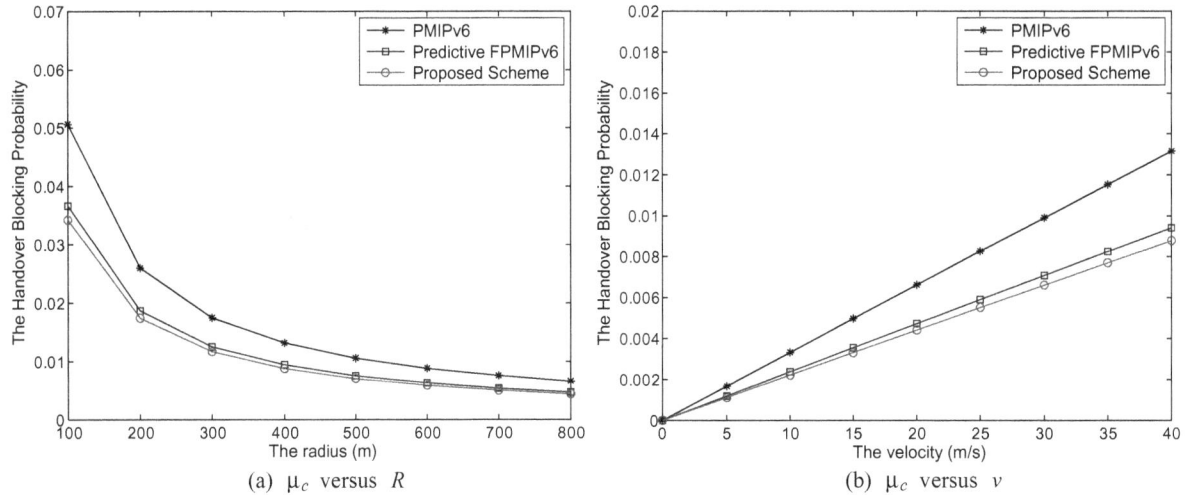

(a) $\mu_c$ versus $R$

(b) $\mu_c$ versus $v$

Fig. 8. The variation of the handover blocking probability.

the variation of the handover latency as a function of $n$ in Fig. 7(b). As the MN performs its handover continuously, the handover latency cumulatively increases and this phenomenon obviously shows that the proposed scheme requires lower handover latency due to its reduced handover authentication time.

In Fig. 8, we investigate the variation of the handover blocking probability. As functions for the variation of the handover blocking probability, we use $R$ and $\nu$. We fix $T_{Auth}$ and $\nu$ as 80 ms and 40 m/s, respectively. Then, we see the variation of the handover blocking probability as a function of $R$ in Fig. 8(a). In the small size of network, the MN quickly moves out to other network so that its handover must be completed in a short time. Accordingly, all schemes show low performance in the small size of network. Next, we vary $\nu$ from 0 to 40 m/s with $R = 400$ m. Then, we see the variation of the handover blocking probability as a function of $\nu$ in Fig. 8(b). Similarly, the MN quickly moves out to other network as its velocity is high. Accordingly, all schemes show low performance in the high velocity environments. From the results presented in Fig. 8, we can confirm that Predictive FPMIPv6 and the proposed authentication scheme provide better performance compare to PMIPv6 due to the reduced handover latency. In addition, the proposed authentication scheme also outperforms others.

## 5. Conclusions

The proposed authentication scheme adopts a variant of the DH key agreement that does not require to have pre-established security associations between relevant MAGs. By avoiding such fixed security associations, the proposed DH key based authentication scheme also improves the scalability of PMIPv6. In addition, the proposed scheme reuses the previously assigned session keys for an MN when the MN changes its point of attachment. Accordingly, in the proposed scheme, the number of authentication and session key generation queries to the AAA server is minimized. We presented the protocol operation and security analysis of the proposed scheme. Then, we have developed the analytical model to investigate the handover latency and the handover blocking probability. The numerical results corroborate that the proposed scheme reduces the handover authentication latency and it outperforms PMIPv6 and Predictive FPMIPv6 in terms of handover latency and handover blocking probability.

## References

[1] A.D.L. Oliva, A. Banchs, I. Soto, T. Melia and A. Vidal, An overview of IEEE 802.21: media-independent handover services, *IEEE Wireless Communications* **15**(4) (2008), 96–103.

[2] A. Durresi, M. Durresi1 and L. Barolli, Secure authentication in heterogeneous wireless networks, *Mobile Information Systems*, **4**(2) (2008), 119–130.

[3] A.M. Hanashi, I. Awan and M. Woodward, Performance evaluation with different mobility models for dynamic probabilistic flooding in MANETs, *Mobile Information Systems* **5**(1) (2009), 65–80.

[4] C. de Laat, G. Gross, L. Gommans, J. Vollbrecht and D. Spence, Generic AAA Architecture, *RFC 2903* (2000).

[5] C. Makaya and S. Pierre, An Analytical Framework for Performance Evaluation of IPv6-Based mobility Management Protocols, *IEEE Transactions on Wireless Communications* **7**(3) (2008), 972–983.

[6] C. Perkins and P. Calhoun, Authentication, Authorization, and Accounting (AAA) Registration Keys for Mobile IPv4, *RFC 3957* (2005).

[7] D. Griffith and R. Rouil and N. Golmie, Performance Metrics for IEEE 802.21 Media Independent Handover (MIH) Signaling, *Wireless Personal Communications*, DOI: 10.1007/s11277-008-9629-4, 2009.

[8] D. Johnson, C. Perkins and J. Arkko, Mobility Support in IPv6, *RFC 3775* (2004).

[9] F. Xia, B. Sarikaya, J. Korhonen, S. Gundavelli and D. Damic, RADIUS Support for Proxy Mobile IPv6, *draft-xia-netlmm-radius-04* (*work in progress*) (2009).

[10] G. Giaretta, J. Kempf and V. Devarapalli, Mobile IPv6 Bootstrapping in Split Scenario, *RFC 5026* (2007).

[11] H. Yokota, K. Chowdhury, R. Koodli, B. Patil and F. Xia, Fast Handovers for Proxy Mobile IPv6, *draft-ietf-mipshop-pfmipv6-09* (*work in progress*) (2009).

[12] J. McNair, I.F. Akyildiz and M.D. Bender, An Inter-System Handoff Technique for the IMT-2000 System, In *Proceedings of IEEE Annual Joint Conference of the IEEE Computer and Communications Societies* (*INFOCOM*) (2000).

[13] J.-H. Lee and T.-M. Chung, Secure Handover for Proxy Mobile IPv6 in Next-Generation Communications: Scenarios and Performance, *Wireless Communications and Mobile Computing*, DOI: 10.1002/wcm.895, 2009.

[14] J.-H. Lee, H.-J. Lim and T.-M. Chung, A competent global mobility support scheme in NETLMM, *International Journal of Electronics and Communications* **63**(11) (2009), 950–967.

[15] J.-H. Lee, Y.-H. Han, S. Gundavelli and T.-M. Chung, A comparative performance analysis on Hierarchical Mobile IPv6 and Proxy Mobile IPv6, *Telecommunication Systems* **41**(4) (2009), 279–292.

[16] J. Korhonen, J. Bournelle, K. Chowdhury and A. Muhanna, Diameter Proxy Mobile IPv6: Mobile Access Gateway and Local Mobility Anchor Interaction with Diameter Server, *draft-ietf-dime-pmip6-04* (*work in progress*), 2009.

[17] K.-S. Kong, W. Lee, Y.-H. Han and M.-K. Shin, Handover Latency Analysis of a Network-Based Localized Mobility Management Protocol, In *Proceedings of IEEE International Conference on Communications* (*ICC*) (2008).

[18] MIPSHOP working group, http://tools.ietf.org/wg/mipshop/, Accssed 2009.

[19]   NETEXT working group, http://tools.ietf.org/wg/netext/, Accssed 2009.
[20]   R. Koodli, Mobile IPv6 Fast Handovers, *RFC 5268* (2008).
[21]   S. Gundavelli, K. Leung, V. Devarapalli, K. Chowdhury and B. Patil, Proxy Mobile IPv6, *RFC 5213* (2008).
[22]   S. Jacobs and S. Belgard, Mobile IP Public Key Based Authentication, *draft-jacobs-mobileip-pki-auth-03* (*work in progress*) (2001).
[23]   S. Yang, H. Zhou, Y. Qin and H. Zhang, SHIP: Cross-layer mobility management scheme based on Session Initiation Protocol and Host Identity Protocol, *Telecommunication Systems*, DOI: 10.1007/s11235-009-9164-y, 2009.

**HyunGon Kim** received the B.S and M.S. degrees in electrical engineering from the KumOh National University, and the Ph.D. degree in computer science from the ChungNam National University, South Korea. He is currently an assistant professor in Department of Information Security at the Mokpo National University, South Korea. He has been with the Electronics and Telecommunications Research Institute (ETRI), Daejeon, South Korea, as a senior member of engineering staff for 11 years. His research interests include wireless sensor network security, telecommunications network security, and vehicle communication security. He has published more than 45 papers in referred journals and conference proceedings. He is a member of IEEE and served as program committee member in several international conferences.

**Jong-Hyouk Lee** received his B.S. degree in Information System Engineering from Daejeon University, Daejeon, Korea in 2004 and his M.S. degree in Computer Engineering at Sungkyunkwan University, Suwon, Korea in 2007. He obtained his Ph.D. degree in Electrical and Computer Engineering at Sungkyunkwan University in 2010. He worked as an intern for IMARA Team, INRIA, France in 2009. He received Excellent Research Awards (two times) from Department of Electrical and Computer Engineering, Sungkyunkwan University. He received Best Paper Award from International Conference on Systems and Networks Communications 2008. Currently, He is a postdoctoral researcher in IMARA Team, INRIA, France. He is now developing a solution to make efficient and secure communications for NEMO based vehicular networks. His research interests include mobility management, security, and performance analysis based on protocol operation for next-generation wireless mobile networks.

# 8

# Glowbal IP: An adaptive and transparent IPv6 integration in the Internet of Things

Antonio J. Jara*, Miguel A. Zamora and Antonio Skarmeta
*Department of Information and Communication Engineering, University of Murcia, Murcia, Spain*

**Abstract.** The Internet of Things (IoT) requires scalability, extensibility and a transparent integration of multi-technology in order to reach an efficient support for global communications, discovery and look-up, as well as access to services and information. To achieve these goals, it is necessary to enable a homogenous and seamless machine-to-machine (M2M) communication mechanism allowing global access to devices, sensors and smart objects. In this respect, the proposed answer to these technological requirements is called Glowbal IP, which is based on a homogeneous access to the devices/sensors offered by the IPv6 addressing and core network. Glowbal IP's main advantages with regard to 6LoWPAN/IPv6 are not only that it presents a low overhead to reach a higher performance on a regular basis, but also that it determines the session and identifies global access by means of a session layer defined over the application layer. Technologies without any native support for IP are thereby adaptable to IP e.g. IEEE 802.15.4 and Bluetooth Low Energy. This extension towards the IPv6 network opens access to the features and methods of the devices through a homogenous access based on WebServices (e.g. RESTFul/CoAP). In addition to this, Glowbal IP offers global interoperability among the different devices, and interoperability with external servers and users applications. All in all, it allows the storage of information related to the devices in the network through the extension of the Domain Name System (DNS) from the IPv6 core network, by adding the Service Directory extension (DNS-SD) to store information about the sensors, their properties and functionality. A step forward in network-based information systems is thereby reached, allowing a homogenous discovery, and access to the devices from the IoT. Thus, the IoT capabilities are exploited by allowing an easier and more transparent integration of the end users applications with sensors for the future evaluations and use cases.

Keywords: Internet of Things, machine to machine, 6LoWPAN, address management, global communications, end-to-end addressability

## 1. Introduction

Flexibility, ubiquity, and scalability are the three required features within the current technological Era, focused on the ubiquitous computing [1] and the Internet of Things communications [2]. Flexibility is required due to the wide range of heterogeneous environments located around the world. The application areas defined usually cover a wide range of domains, including user-centric solutions such as e-health [3]; solutions more indirectly related to the user such as environmental monitoring [4]; those totally focused on the user, like Intelligent Transport Systems [5]; and finally, the development of a mix of solutions not focused on the user combined with user application domains, such as smart cities [6]. These solutions can range from simple sensors under the road for parking areas, or in the floor to measure humidity, to more complex systems, such as smart meters to measure pollution, air quality, and environmental

---

*Corresponding author: Antonio J. Jara, Department of Information and Communication Engineering, University of Murcia, Murcia, Spain. E-mail: jara@um.es.

factors, and can even include clinical devices in e-Health and solutions for Ambient Assisted Living environments [7].

Flexibility, ubiquity and scalability are properties found in the current Internet, and that is why the aforementioned challenges can be solved not only with the new capabilities to link Internet with everyday sensors and devices, but also with the exploitation of data captured from the Future Internet through the so-called Internet of Things (IoT).

Future Internet and the IoT represent unprecedented growth in the number of devices and users connected to the Internet. Therefore, the devices should be as autonomous as possible in satisfying the so-called 'self-* functionalities', such as self-management, self-healing, and self-discovery. These properties are especially challenging in the Internet of Things, where many devices are mobile and, consequently, can change their location in the network.

For that reason, this research is focused on operating on top of the Future Internet infrastructure, i.e. IPv6. Thereby, users and clients discover and use homogenous IP-based resources, with protocols and technologies that are very well-known and are already deployed. However, other technologies such as IEEE 802.15.4, and Bluetooth Low Energy, are not based on IP networks, and have different limitations. For those technologies, other uses out-of-IPv6-network mechanisms have been defined: device location is not performed with IP locators, discovery messages are carried over a path not intended for general data communication, and so on.

As previously mentioned, our principle goal is to avoid the out-of-IPv6-network mechanisms in order to homogenize the discovery and use of resources through the Future Internet infrastructure, i.e. through the IPv6 network. This makes services reachable through homogenous and interoperable technologies, such as WebServices, and the discovery of services can be conducted through network-based Information Systems that are already deployed, such as the Domain Name System with the Service Discovery extension (DNS-SD).

The contribution of this paper is a way to provide any sensor, device or platform with IoT capabilities, from the common networking point of view. In other words, this means the evolution of consumer devices, as well as communication with the capabilities offered by the Future Internet, based on IPv6 protocol and technologies, such as IPv6 over Low Power Area Networks (6LoWPAN). In this regard, 6LoWPAN allows Internet extension to small and smart devices, making it feasible to identify and create connections among people, devices, and the things surrounding us.

Consequently, this extension to Internet, based on IPv6 and global connectivity, offers a homogeneous support and end-to-end capabilities. Furthermore, the technologies and current technologies/protocols from 6LoWPAN and IPv6 can be extended and re-used, e.g.: directory systems based on Domain Name System Service Discovery (DNS-SD), which can extend the current Resource Directory from M2M platforms and projects, such as SENSEI [8], in a global way.

Since original IPv6 and 6LoWPAN are not extensively supported by sensors and devices, such as legacy devices, nor by the environmental sensors currently deployed, this work instead proposes a mechanism for efficient support of global communications in machine-to-machine (M2M) scenarios, called Glowbal IP.

Glowbal IP is the result of scientific research on the two main open issues found in Internet of Things scenarios with applications and use-cases, oriented to connectivity with external servers, mobile user applications and interoperability scenarios.

The first concern is that most of the sensors and technologies currently deployed in smart cities, and industrial, building automation and smart grid scenarios lack 6LoWPAN support and IPv6 capabilities. As a result, current technologies must be extended to enable IPv6 capabilities, either with new hardware or with dramatic changes in the communication stack. These options are usually neither feasible nor scalable, and are not cost-effective for big deployments where it is not supported maintenance mechanisms such as Over-the-Air (OTA) firmware updates. Therefore, the main design challenge for Glowbal IP was to define a solution to be laid over the existing application layer, in order to avoid any changes in the current communication stacks, or any need for additional capabilities. Glowbal IP will be a part of the application to extend the addressability and determine end-to-end sessions for these technologies and devices.

A second problem was found with 6LoWPAN performance, which is the reference technology to integrate IPv6 in smart devices with limited capabilities due to its ability to link the Internet with everyday devices. In general terms, we found that the RFC4994 [9] overload was too high for global communications, since it requires in-line consideration of the full address; however, changes made by the new RFC6282 [10] to minimize the problem were insufficient, since it still requires a complex context management for an overload reduction that is not significant. We therefore concluded that even the technologies currently supporting IPv6 for smart things offer less than optimal header compression.

As a result, 6LoWPAN devices prepared to optimize performance within global communication, as well as other smart devices, can only be powered with 6LoWPAN or native IPv6. This work, however, proposes a new adaptation protocol called "Glowbal IP" which provides an Access Address Identifier (AAID), and an AAID-IPv6 address translation mechanism for different technologies, in order to adapt any device to the Internet of Things architecture via IPv6. In this respect, AAID simplifies all the parameters from IPv6 communications (source address, destination address, source port and destination port) in a single 4-byte communication identifier. Thus, the mentioned IPv6/UDP header overload that occurs in devices based on 6LoWPAN can be reduced with global IP addressing.

Therefore, Glowbal IP reaches an effective frame format for global communications, and a mechanism to support global communications with the Future Internet architecture, i.e.: with other IPv6 end nodes in networks that do not support IPv6 or 6LoWPAN. The main advantages from Glowbal IP are the reduction of the overhead of 6LowPAN networks for global communications, and the integration of legacy systems that cannot implement IP protocols due to various issues such as limited payload size e.g. Bluetooth Low Energy, or because it was not considered during the deployment, therefore it is offered a solution adaptable and transparent to the exiting deployments and communication stacks.

This document is organized as follows: Section 2 presents related works that deal with aspects such as IPv6 and 6LoWPAN, as well as their limitations, and the capabilities to define network-based information systems to discover resources and services by means of the Domain Name System Service Directory; Section 3 describes the Glowbal IP protocol, with the different format and entities involved in the mapping; Section 4 demonstrates the advantages of Glowbal IP to enable network-based Information Systems through DNS-SD and WebServices; Section 5 describes how the protocol has been implemented and evaluated; Section 6 explains Glowbal IP's advantages and capabilities; and Section 7 offers conclusions and future lines of research.

## 2. Related works

This section describes the works related to Glowbal IP from two perspectives. First, it describes the advantages of integrating IPv6 in the Internet of Things, as well as the limitations from current approaches, i.e. 6LoWPAN. Second, it describes the capabilities to build advanced network-based information systems, such as resource and services directories, over the Future Internet architecture, by using technologies such as DNS-SD.

*2.1. Future Internet (IPv6) and the Internet of Things*

The Internet of Things (IoT) is the main justification for the Future Internet, where IPv6 is the fundamental technology. It is estimated that the Future Internet will reach from 50 to 100 billion connected things by 2020 [11], while the IPv6 address space supports $2^{128}$ unique addresses (approximately $3.4 \times 10^{38}$); specifically, it can offer $1.7 \times 10^{17}$ addresses on an area about the size of the tip of your pen. As a result, it makes it feasible for sensors and consumer devices to evolve towards the communication capabilities presented by the Future Internet with IPv6 protocol, by means of new technologies. It can be found several stacks and implementations for the integration of IP in wireless sensor networks [12], but the most extended and standardized is IPv6 over Low Power Area Networks (6LoWPAN). This extends the Internet to small and smart devices. Therefore, it will allow systems to identify and connect people, devices, and the things around us.

The 6LoWPAN adaptation layer was indicated to carry IPv6 datagrams over constrained links, such as the one defined in the IEEE 802.15.4 standard [13], taking into account the limited bandwidth, memory, or energy resources that are expected in applications such as wireless sensor networks [14]. IEEE 802.15.4 represents a Maximum Transfer Unit (MTU) of 127 bytes. This is reduced to 102 bytes after taking into account the IEEE 802.15.4 MAC header with extended MAC addressing IEEE EUI-64bits. It is further reduced by an additional 21 bytes [15] after accounting for link layer security (AES-CCM-128 for integrity and confidentiality). This leaves from 81 to 102 bytes of actual payload, depending on security level.

6LoWPAN defines a Mesh Addressing header to support sub-IP forwarding, a Fragmentation header to support the IPv6 minimum MTU requirement, and stateless header compression for IPv6 datagrams. Specifically, the defined 6LoWPAN headers, one for the IP header, and another one for the UDP header (LOWPAN_HC1 and LOWPAN_HC2, respectively) are able to reduce the relatively large IPv6 and UDP headers down to several bytes.

LOWPAN_HC1 and LOWPAN_HC2 are insufficient for most practical uses of IPv6 in 6LoWPANs, where global communications are involved [16,17]. LOWPAN_HC1 is most effective for link-local unicast communications, where IPv6 addresses carry the link-local prefix and an Interface Identifier (IID) directly derived from IEEE 802.15.4 addresses. In this case, both addresses may be completely elided. However, even though link-local addresses are commonly used for local protocol interactions, such as IPv6 Neighbor Discovery, DHCPv6, or routing protocols, they are not commonly used for the application layer, where communication with external clients and servers is required.

To solve the problem found with RFC4944 [9], a new encoding format, LOWPAN_IPHC, was developed for RFC6284 [10] for effective compression of Unique Local, Global, and multicast IPv6. This new encoding format is based on shared state within contexts. But, although usable, this is still inefficient. Figure 1 presents stack overhead for the RFC4944 and for the RFC6282 with contexts.

Figure 1 shows that 6LoWPAN has an overload of 26–41 bytes, meaning that the final available payload is reduced to half of the original size, i.e. 61 to 76 bytes from the original 127 bytes. Therefore, it is reduced to less than 50% of the original frame size from IEEE 802.15.4 of 125 bytes.

*2.2. Directory systems for the Internet of Things*

Two different levels of discovery can be defined: *resource discovery*, the discovery of devices on the network; and *service discovery*, the discovery of the services, methods and functions offered by a specific resource.

| IP Protocol Stacks | | | | IoT Protocol Stacks with 6LoWPAN | | |
|---|---|---|---|---|---|---|
| HTTP | RTP | | Application | YOAPY | Application protocols | Payload available ¯6 OR *61 bytes* respectively |
| TCP | UDP | ICMP | Transport | UDP | ICMP | *26 bytes* with global address based on contexts OR *41 bytes* with address in-line |
| IP | | | Network | IPv6 | | |
| | | | | LoWPAN | | |
| Ethernet MAC | | | Data Link | IEEE 802.15.4 MAC | | *25 bytes* with extended MAC address based on IEEE EUI-64 |
| Ethernet PHY | | | Physical | IEEE 802.15.4 PHY | | |

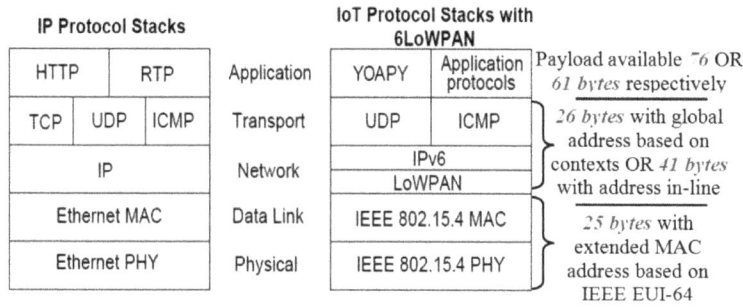

Fig. 1. Future Internet and Internet of Things stacks.

Resources are reachable through technologies such as 6LoWPAN, Bluetooth Low Energy and, from our point of view, any technology offering IPv6 support. At least when they are not reachable, they are identifiable through technologies such as Radio Frequency Identification (RFID), and Near Field Communication (NFC).

*Resource discovery* is the process by which the user is able to find devices offering services according to his criteria and interests. It can differ from the resources that the user can explicitly request or from a more sophisticated discovery where the network is more pro-active, and it notifies the user about the availability of these new devices [18].

Resource discovery will provide descriptive information, such as the resource type or family, and some attributes to describe it. In addition, it will provide the information that the user will need to reach them, i.e.: a locator such as a URL or IP address.

Resource discovery management requires dynamic updates to the system with the new resources included in the network, as well as the ability to integrate the updates over mobile [19] in order to be consistent with the real resources reachable at a specific moment, and security [20] to protect resources and services that one may want to access through controls.

*Service discovery* is focused on the description of those services provided by technologies, such as those that are Web-based, i.e.: XML, Web Services, or other technologies such as JSON, and DNS Service Directory.

These services include printing and file transfer, music sharing, servers for pictures, documents and other file sharing, as well as services provided by other resources. With the expansion towards the Internet of Things, other, simpler services can be considered, such as the environmental status consultation for temperature, humidity and lighting, a pressure value for a parking sensor, or the value of glucose for a medical device.

Different techniques can be found for resource and services discovery and the Internet of Things. Right now, the most common approach is the definition of M2M platforms [21], such as ThingWorx, Pachube, Sen.Se, and SENSEI, where the devices are registered in the platform, and are reachable from the Internet through WebServices such as SOAP and REST. The problem with this static approach is that it is limited to the information on the platform and the manually registered devices and, for that reason, defining more scalable solutions must be required. This makes it possible for resources entering the network to be available by registering with the discovery system, without any interaction with the user, on a directory system that is homogenous and consultable simply over the Internet, without the need to use a specific M2M platform.

This capability for autonomous registration and the discovery functionality to be dynamically adapted with the inclusion of new devices in the network is necessary for the above-mentioned IoT to be flexible and ubiquitous. It is also necessary for the scalability required to manage every resource connected to the network, whose number is continuously increasing. It is not feasible to continue considering IoT solutions that require a manual and static management of resources, with fixed registration over specific directory systems from the M2M platforms.

Some naming systems such as Lightweight Directory Access Protocol (LDAP) [22], Universal Description, Discovery, and Integration (UDDI) [23], and Domain Name System (DNS) [24] offer resource and service directory capabilities, and more specific resource discovery technologies could be added, such as UPnP, JINI, Service Location Protocol (SLP), and Rendezvous or Bonjour protocol over DNS-SD.

The solution considered for the Internet of Things has been DNS-SD because an independent directory for the resource is needed. Consequently, other approaches based on multicast queries such as SLP – with direct discovery of resources – are not feasible for the IoT. That solution does not work properly because, for one, the resources are not discoverable while they are sleeping, and they need sleep in order to optimize their power consumption and lifetime. It also does not work properly because the overload from multicast for this kind of mesh topology network is not appropriate and, finally, because continuous requests to the node overloads the end-device.

Rendezvous protocol from ZeroConf Architecture, which is used for Bonjour in MAC OS, or AVAHI in Linux OS are based on the DNS-SD from the IETF ZeroConf working group [25] to store the information and multicast DNS (mDNS) in order to query the DNS-SD records.

DNS-SD and mDNS present a solution where no additional infrastructure is needed, and merely requires that resources be enabled with an IP address. The solution is focused on the re-use and extension of existing Internet standards. This can be found with a multicast approach, such as the first stage Service Location Protocol and JINI protocols through multicast, particularly multicast DNS (mDNS), and the DNS service discovery (DNS-SD) to leverage existing Internet protocols. In our approach, following the objective from Glowbal IP to enable all resources with IPv6 addresses, and the re-use and extension of current Internet technologies, we will focus on DNS-SD, as we will describe in Section 4.

Section 4 will also describe the use of common Web technologies for the definition of services, the same way UPnP leverages common Web technologies such as HTTP, SOAP, and XML, to provide access to resources and services. More specifically, WebServices is used for the Internet of Things [26]. It is based on RESTFul and CoAP [27], which is a lightweight version of REST, and the description of services in the DNS-SD [28] follows the semantic [29] and naming conventions that describe how services will be represented in DNS records, as defined by Web Linking description, in particular the version of Link format defined under the CoRE IETF working group [30].

## 3. Glowbal IP protocol

Taking into account not only IoT heterogeneity, but also the facts that 6LoWPAN (RFC4944 and RFC6282 version) represents low optimization for global communications and that several technologies do not offer native IPv6 support or 6LoWPAN, a more optimized adaptation must be considered. This adaptation for global communications should be compatible with IPv6 and 6LoWPAN, allowing IPv6 to connect to networks with non-IP support, such as Bluetooth and LoWPANs, with only IEEE 802.15.4.

The protocol proposed here provides for an Access Address/Identifier (AAID) which simplifies every parameter from IPv6 communications (source and destination address/port, 36 bytes) in a single 4-byte communication identifier. Thus, the 41 bytes from the IPv6/UDP headers based on 6LoWPAN with global IP addressing can be significantly reduced. This achieves an effective frame format for global communications, and also a mechanism to support global communications in networks that do not have native support for IPv6.

This section will describe the Glowbal IP protocol, its motivation and objectives, the implementation details with its header format description, the tables involved in mapping between Glowbal IP identifiers and IPv6 addresses, and finally some examples of how it can enable IP connectivity to allow interoperability with the rest of the IPv6-based devices connected to the Future Internet Core Network.

### 3.1. Motivation and objectives

The main motivations for defining the Glowbal IP protocol have been the overload resulting from 6LoWPAN for global communications, as stated above in Section 2.1. Another motive has been to make IP connectivity feasible for non-IP enabled technologies and new IoT technologies, such as Bluetooth Low Energy (BT-LE), which has a limited payload of 19 to 27 bytes.

Based on those motivations, Glowbal IP's objective is to provide connectivity to the Future Internet Core Network based on IPv6 for devices such as IEEE 802.15.4 sensors, which have no support for 6LoWPAN in their stacks. In addition, the purpose is to provide connectivity to other technologies which do not provide any global communication capability in their stack, although they are able to communicate with the Core Network through other interfaces located on their platform. Bluetooth, for example, can be located through handset devices. These handheld devices offer connectivity with the Internet through their GPRS/GSM network interfaces. Therefore, they can be used as a gateway to manage a personal network, where each device is connected to a smart phone through Bluetooth Low Energy, Bluetooth 2.1 or Near Field Communication (NFC) by means of its active mode NDEF Push Protocol (NPP) [31] that can be addressed and can communicate globally. This work is focused on networks based on IEEE 802.15.4, but ongoing work will focus on extending Glowbal IP for Bluetooth Low Energy as well.

Glowbal IP technologically provides current sensors from deployed environments – where IPv6 is not provided – with an adaptation mechanism to make end-to-end connectivity feasible.

Furthermore, it allows interoperability among different scenarios and technologies deployed providing, in turn, an easier mechanism based on the Internet that helps integrate sensors from external providers through a transparent IPv6-based reach. In addition, to these integration heterogeneous multi domain networks [32], it will be addressed in the future issues such as multi-homing, and mobility, which are also some capabilities from IP technologies [20,33–35].

### 3.2. AAID: Access Address Identifier

Access Address/Identifier (AAID) is a 32-bit address that is generated at the time the connection is set up. The access address identifies a connection between a node and a client.

The sensors can configure/negotiate the AAID with the gateways/Border Routers for a communication process. The usual overload derived from IPv6 address size occurs only at initialization. After initialization, AAID simplifies the parameters from the IPv6 communication (source address, destination address, source port and destination port) in a single 32-bit communication identifier (4 bytes). This reduces the IPv6/UDP headers overload based on IPv6, and even 6LoWPAN with global IP addressing.

In order to make this process more compatible and extendible, AAID has not been considered as an additional network layer; instead, it has been included as part of the current application layer (payload), making this end-to-end connectivity transparent for current end-nodes, and for the communications stacks already defined.

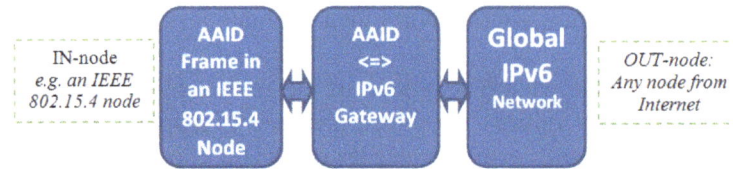

Fig. 2. AAID translation from Intranet of Things to Internet of Things.

Translation is carried out by the gateways, as shown in Fig. 2.

As we can see, global connectivity is reached through the AAID to IPv6 gateway, which is called an AAID gateway. This AAID gateway's features are similar to the Border Routers from 6LoWPAN, where the Border Router also performs the adaptations in a similar way as the method defined for Glowbal IP.

Finally, with regard to the AAID generation mechanism, a simple hash process is defined based on CRC-32 bits, since this offers low complexity. It can be automatically calculated by the end-node and by the AAID gateway. The hashing is calculated over the session information, as it is defined in Eq. (1).

$$AAID = h(\text{source IP, destination IP, destination Port}) \tag{1}$$

This AAID value belongs to each node, and is used for the session only. Therefore, it does not presenting aliasing problems, as described in the following section for the Local to Global mapping tables (L2G).

The complexity of AAID generation and management lies in supporting multi-homing and mobility where a different AAID is generated for each of the different gateways, since they assign a different IP source for this AAID node. Therefore, a synchronization mechanism is required for all the AAID gateways involved.

For mobility purpose, it is mainly required a synchronization of the previous AAIDs established in the previous AAID gateway, and the new visited AAID gateway, in order to avoid recalculate them for the new IP address. Therefore, it will be required the transfer and synchronization of the AAID records and information associated in the L2G tables during the binding transfer as part of the mobility protocol.

Mobility and multi-homing support is currently being considered for Glowbal IP, and the required fields in the header format and dispatch have been defined in order to report this situation. More specifically, however, this extension of Glowbal IP for mobility and multi-homing support is part of the ongoing work.

### 3.3. Glowbal IP header format

Glowbal IP is not actually presenting a network header since it is considered part of the payload. Therefore, it can be considered as a session layer over the application layer.

As it has been mentioned, the IPv6 source is generated through mechanisms such as stateless auto-configuration. Therefore, explicit indication is not necessary.

Regarding source port, it can be found two options; on the one hand, it can be considered that each device has a unique application; in this case it is not necessary to indicate the source port either. However, on the other hand, it can considered different applications, such as ports for management applications, e.g. Simple Network Management Protocol (SNMP), control applications for supervisor or privileged use access, and finally user applications. In addition, this should not be limited the capabilities of applications in the future in order to offer a high flexibility for the Internet of Things. For that reason, Glowbal IP offers both options, in the case of working with ports, it is generated an AAID for each one

| | | |
|---|---|---|
| S | Set Mode (1 bit) | Enabled when setting AAID values |
| M | Mobility (1 bit) | Enabled when the node is mobile |
| IP src | Source IP configuration (2 bits) | 00-Autoconfiguration EUI-64 (0 bytes) / 01- in-line host id address (8 bytes) / 10- in-line full address (16 bytes) / 11- Generated by GW (0 bytes) |
| IP dst | Set Destination IP (1 bit) | Enabled when the destination IP is in-line for setting up or update the value |
| Port src | Source Port configuration (2 bits) | 00- Default Port or defined by GW (0 bytes) / 01- Short subrange from 6LoWPAN (1 byte) / 10- Long subrange from 6LoWPAN (2 bytes) / 11-in-line full port (4 bytes) |
| Port dst | Set Destination Port (1 bit) | Enabled when the destination Port is in-line for setting up or update the value |

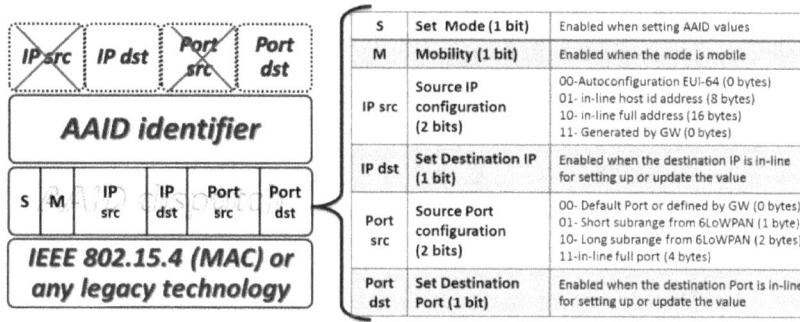

Fig. 3. AAID format and meaning fields.

of the ports, since it is part of the AAID generation mechanism, see Eq. (1). Finally, regarding security, it is highly interesting in order to manage policies, where specific ports can be reserved for management, and consequently controlled to now allow access from external network, allow only access to specific hosts, or apply an advanced control access.

Right now, the applications are moving around the denominated Web of Things, where the Internet of Things nodes are embedded Web Services server, which deals with all queries regarding the announcement of CoAP methods.

Glowbal IP offers the option of indicating the IPv6 source, in order to be more flexible and for specific use-cases where an end-node uses static addressing is required.

AAID requires information regarding the IP network layer, mainly for the destination node. These factors about specification of source information will simplify AAID management, particularly during mobility and multi-homing.

The IPv6 header format consists of three main parts:

– *AAID dispatch:* Indicates the control information for AAID management, as well as information from the original IP/UDP header in-line. Specifically, the fields provided are:

* *Set bit (S):* This bit indicates the definition of a new session from the AAID gateway to the AAID node, or vice versa. It is usually accompanied by a set of in-line fields, such as destination port, destination IPv6 address, etc.
* *Mobility/Multi-homing bit (M):* This tells the AAID gateway to check with the other AAID gateways, through the back end or the overlay, that this node does not have pending sessions from another location.
* *IP source format (IP S):* These bits indicate whether the IP source address is based on auto-configuration through the mechanism defined in the RFC 4862 (00); whether it requires the use of a specific host id (01) to keep a fixed address in a local domain; whether it is a state-full address, regularly used by this node (10); or whether it has been delegated to the AAID gateway (11) to use any of the mechanisms described in Section 3.4 for generating the source IP address in the L2G table.
* *IP destination set (IP D):* This bit shows its configuration or lack thereof. This function is usually enabled in conjunction with the S bit.
* *Port source format (Port S):* This is based on the same format as the one for 6LoWPAN in RFC4944, where specific sub-ranges of ports are defined in order to reduce the number of bytes that are required to specify the source port. In addition, as already mentioned, the determination of the port from the AAID gateway can be taken into consideration to simplify the process.

∗ *Port destination set (Port D):* This bit shows its configuration of its lack of it. This function is usually enabled in conjunction with S bit

- *AAID identifier:* This defines the 32 bits of the AAID identifier.
- *In-line fields:* Finally, the fields which are in-line for the setting phase of the AAID are closed. We must underscore that the in-fields are only done for the initialization packet, while the rest of the session only performs the dispatch and AAID. In other words, this means only 5 bytes will identify all communications, instead of the usual 26 to 41 bytes from 6LoWPAN.

### 3.4. Link Layer to Global IPv6 Mapping Table (L2G Table)

In addition to the foregoing issues regarding AAID management with multiple gateways, there is another crucial goal to be met: maintaining uniqueness across the gateways for an AAID association. To this end, the Link Layer to Global IPv6 mapping Table (the L2G table) defines a relationship between the AAID and the MAC address from the device. The only requirement is to guarantee the AAID uniqueness inside a specific device; in the event that an alias conflict (aliasing) takes place between two sessions from the same device, a simple solution can be applied to avoid it, such as a plus one mechanism, i.e. adding one unit to the AAID number when there is some conflict with a previous AAID.

This potential conflict does not require more complex management, since both the AAID gateway and the AAID sensor are aware of the current AAID node sessions, even in scenarios with mobility and multi-homing.

The L2G table stores the mapping between the Link Local address based on MAC EUI-64, the short address in the case of IEEE 802.15.4, and the assigned IPv6 global address.

In the specific case of IEEE 802.15.4, the mapping process can be automated by means of the stateless auto-configuration defined in the RFC4862 [36], where the Interface Identifier for IEEE 802.15.4, defined in the RFC4291 [37], is based on the MAC EUI-64 identifier assigned to the device. The global address for the end device is then created with the combination of the IPv6 prefix and the EUI-64 address. This is equivalent to the mechanism used to build addresses through the Neighbor Discovery protocol, where the Router Advertisement reports the IPv6 prefix, default router address, hop limit, and MTU.

It is assumed that the communication between the AAID node and the AAID gateway is based on extended link layer addressing, i.e. MAC EUI-64 addressing. However, as for the presented for MAC EUI-64 addressing, it can be also considered the mapping with the 16-bit short address. In this situation, a "pseudo 48-bit address" should be built, as indicated in RFC4944, and extended to EUI-64 with the procedure from RFC2464 [38].

Once the IPv6 address has been built for the AAID node, the AAID for each session can be calculated following Eq. (1).

Figure 4 is an example of the L2G table located in the AAID gateway. This scenario functions as a bridge between the adapted network based on IEEE 802.15.4, and the global IPv6-based network.

As we can see, only the native addressing, the calculated source IP, the destination IP address and destination port for the session are indicated for the L2G table, together with the assigned AAID identifier at the end. Section 5 describes AAID gateway implementation and evaluation of the AAID mapping performance.

### 3.5. Interoperability scenario

IoT consists of a wide range of different scenarios with significant technological differences that prevent direct interoperability. Bearing in mind the foundations of the Future Internet and the Internet

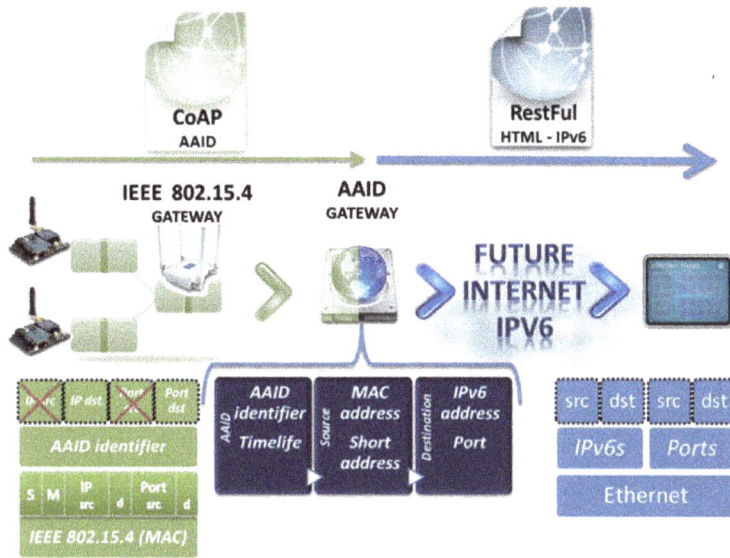

Fig. 4. L2G table for mapping between AAID-based network and IPv6-based network.

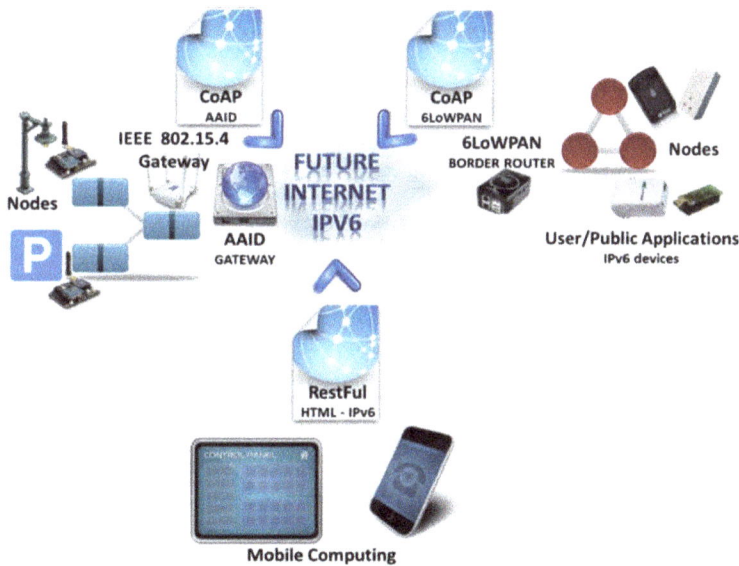

Fig. 5. Inter-operability scenario among Glowbal IP, IPv6 and 6LoWPAN.

of Things, it is therefore imperative to achieve multi-scenario interoperability. This characteristic is one of the advantages offered by the Glowbal IP protocol proposed here.

Figure 5 shows an interoperability scenario, where there are sensors supporting 6LoWPAN and RestFul through Contiki OS, such as Telos B motes, as well as sensors with native IPv6 support, such as SunSpots. It is connected with applications from the user or public management as well, and accessed through the Future Internet network in a transparent and homogeneous way.

The rest of the platforms/technologies/machines that do not offer direct IPv6 or 6LoWPAN support can be adapted in a similar way with AAID over their native application layer (payload). The sensors

| Core Network | 6LoWPAN | GLowbal IP | |
|---|---|---|---|
| HTTP | CoAP | CoAP / HTTP / * | *Request/Response layer for resource manipulation* |
| TCP | LOWPAN_HC2 (UDP) | AAID (IP + UDP) | *Network Transaction Layer* |
| IP | LOWPAN_HC1 (IP) | | |
| (e.g. Ethernet) | IEEE 802.15.4 | IEEE 802.15.4 | |

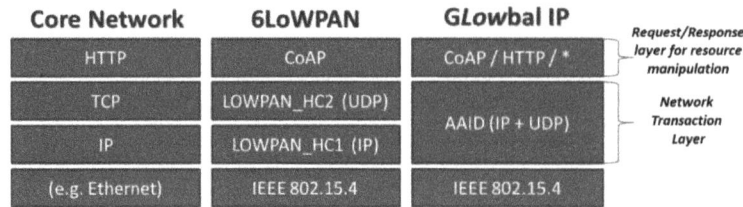

Fig. 6. Different WebServices and application stacks over 6LoWPAN, native IPv6 and Glowbal IP.

considered are the Waspmote nodes, which are highly extended in smart cities and environmental monitoring deployments. Waspmotes only support IEEE 802.15.4. But, these nodes are able to reach IP connectivity through Glowbal IP.

In addition, it is not required to reprogram the communication stack from the nodes, such as it is required to add 6LowPAN support. Thus, Glowbal IP is very useful for deployments where the nodes can be only be reprogrammed OTA for the application layer software, such as it is found in some deployments based on Internet of Things for smart cities such as SmartSantander. Thereby, Glowbal IP is defined in the applications embedded inside the nodes, which is more feasible, scalable and safes that reprogramming the nodes with a full communications stack.

Finally, Glowbal IP allows interoperability with the Future Internet core network, and consequently enables interoperability among scenarios with IPv6 technologies.

## 4. Enabling network-based Information Systems with Glowbal IP

### 4.1. WebServices support with Glowbal IP

Homogenous access to information and management by way of WebServices such as RESTFul/CoAP and SOAP lightweight is highly interesting and desirable in order to define solutions that are based on the so-called Web of Things [39].

WebServices protocols, such as RESTful and other application packets, are encapsulated after AAID, which only identifies the session to which the packet belongs. Therefore, it also opens an opportunity for the Web of Things and remote Web Services with the end-node. The next table shows the adaptation and homogeneity among IP, 6LoWPAN and AAID, providing the upper layers with connectivity support.

Subsequently, for transactions and transport, HTTP over TCP could be applied in the current Internet architecture; CoAP over 6LoWPAN in the current Internet of Things, based on 6LoWPAN; and finally, the proposed adaptation for non-IPv6 networks, i.e. CoAP, HTTP or any protocol over the Access Address/Identification (AAID), which, as it has been mentioned, encapsulates UDP information (porting) and IP information (addressing).

Figure 6 shows the relationships among the different stacks for the network transaction layer, which involves networking and transport, as explained in previous sections. However, it also takes into account the application layer for the request/response of the different resources, as well as services from resources. For this purpose, the most common technology is the Web of Things, which defines the use of REST for the Core Network, and a light version of REST for 6LoWPAN, called CoAP. In the case of Glowbal IP, any of them can be considered since it is independent from the network layer, although CoAP is more frequent due to its advantages for constrained devices.

To do so, the sensors are extended with RESTFul methods and attributes. Web Services offers interoperability and homogenous access to services, regardless of the technology from the underlays.

**DNS Records Look Up tool**

| Domain | Type | Class | Result |
|--------|------|-------|--------|
| light3.rd.esiot.com. | TXT | IN | "rt=light\;ins=3\;lt=86400\;model=normal\;if=802.15.4\;value\;onoff" |
| rd.esiot.com. | NS | IN | rd.esiot.com. |
| rd.esiot.com. | A | IN | 155.54.210.159 |
| rd.esiot.com. | AAAA | IN | 2001:720:1710:0:216:3eff:fe00:9 |

Fig. 7. DNS-SD for a thing based on IEEE 802.15.4 accessed through Glowbal IP.[1]

**DNS Records Look Up tool**

| Domain | Type | Class | Result |
|--------|------|-------|--------|
| light1.rd.esiot.com. | TXT | IN | "rt=light\;ins=1\;lt=86400\;model=normal\;if=X10\;housecode=A\;unitcode=5\;value\;onoff" |
| rd.esiot.com. | NS | IN | rd.esiot.com. |
| rd.esiot.com. | A | IN | 155.54.210.159 |
| rd.esiot.com. | AAAA | IN | 2001:720:1710:0:216:3eff:fe00:9 |

| Domain | Type | Class | Result |
|--------|------|-------|--------|
| light2.rd.esiot.com. | TXT | IN | "rt=light\;ins=2\;lt=86400\;model=dimmer\;if=EIB\;area=1\;zone=2\;deviceID=3\;value\;onoff" |
| rd.esiot.com. | NS | IN | rd.esiot.com. |
| rd.esiot.com. | A | IN | 155.54.210.159 |
| rd.esiot.com. | AAAA | IN | 2001:720:1710:0:216:3eff:fe00:9 |

Fig. 8. DNS-SD for a thing based on X10 (top) and EIB/KNX (bottom) accessed through Glowbal IP.

For example, the same set of sensors is considered with different technologies, even when the physical sensor comes from a different manufacturer, such as IEEE 802.15.4, X10 and EIB/KNX. For example, a common set of RESTFul services interface has been defined for the attributes of a light. This will prove scenario independence, since both can be accessed at the same way, by defining the appropriate IPv6 address. *See* next section, and Figs 7 and 8, where all the light sensors offer the same "onoff" and "status" CoAP methods in order to change the status, and then consult the respective status.

## 4.2. Discovery support with Glowbal IP

Discovery allows resources to become aware of services from the rest of the resources without explicit management from the user. This means that a user, and even another resource, can discover and potentially use a resource without prior knowledge of it, and of its capabilities and services.

Section 2.2 discussed a wide range of discovery systems, most of which are in use today. It concluded that the majority of them look for a homogenous way to locate devices, and the solution used to locate resources is IP. We should emphasize that IP addressing is not directly used on a regular basis; the URL or device name is offered instead, although, at the end, it is an IP-based locator. Therefore, Glowbal IP presents the advantage of continuing to use the already extended discovery systems for the Internet of Things.

The chosen solution for discovery is DNS-SD and mDNS, which is an evolution of DNS. It is mainly focused on the naming aspects for resources, although it also offers the description of the resource and services through the use of its records. Note that this solution defines neither new operations nor new DNS record types. Therefore, it is fully compatible with the current DNS deployment.

Originally, DNS-SD defines DNS pointer (PTR) records to indicate the type of service using the form _type._protocol.domain for a specific IP locator. These pointers are originally defined by the reverse DNS

---

[1]The different TXT records can be tested for light1.rd.esiot.com, light2.rd.esiot.com, and light3.rd.esiot.com through the DNS tool: http://www.hscripts.com/tools/HDNT/dns-record.php.

Table 1
Link Format and DNS-SD description for the resource

| Link format | DNS-SD |
|---|---|
| Resource Instance (ins=) | {instance} |
| Resource Type {rt=} | {ServiceType} |
| <uri> (It is already obtained from the AAAA entry) | TXT path= |
| Interface Description {if=} | TXT if= |
| Additional attributes (e.g. the CoAP methods){xxx=} | TXT xxx= |

protocol. Through these reverse DNS records, the user is able to find all resources in a specific domain, e.g.: "example.com," by issuing a query for _http._tcp.example.com.

In this case, _http denotes the application-layer protocol that the user is looking for services based on HTTP, while _tcp indicates that the service runs over TCP/IP stack, and example.com denotes the domain to which the query is directed. Furthermore, DNS-SD is extended with mDNS in order to carry out the queries, and use the ".local." domain to operate over link-local multicast.

The aforementioned pointer records contain the service instance names in a format for matching the query. For example, a query for printers might return the URL for the printer available in the domain. DNS-SD overloads this request so that when the client passes a service type in the query, the query returns service instance names.

In addition to the reverse directory, the type of service can be determined, where RFC2872 [40] defines a group of very well-known services, and "TXT" records which will be associated with that service to define the attributes or keys, such as model, ID, type, status, or more specific information for each service.

With regard to the Internet of Things, a simplification of this DNS-SD through the CoRE IETF working group has been defined. For example, this simplification only considers the discover resources by requesting "/.well-known/core".

From the CoRE point of view, the DNS-SD can be seen as a repository for Web Links. The use of Web Linking for description and discovery of resources hosted by constrained web servers is specified by the CoRE Link Format.

Discovery is performed by sending a native query to the DNS-SD through the mDNS protocol, or it can be used through the CoAP interface, where parameters such as Resource Type (rt) are defined to filter the kind of device. It should be pointed out that the functionality through DNS-SD native protocol and CoAP is equivalent, but the CoAP extension is also considered in order to make the discovery process simple and homogenous for devices with constrained capabilities to implement an additional protocol for the DNS.

Specifically, the attributes for a specific device are defined using the well-known CoRE Link Format interface or any other description format over one "TXT" record.

In particular, the CoRE Link Format defines the following attributes for the "TXT" record: Resource Type "rt", Instance attribute "ins", assuming that we can find multiple instances of the same resource type in an environment, e.g.: multiple lights and temperature sensors can frequently be found. In addition, other parameters are considered, such as Interface "if" in case multiple technologies are available on the platform, and other parameters such as lifetime "lt", domain "d", and context "con".

Table 1 presents the differences between the CoRE Link Format used for the Internet of Things and the original DNS-SD. As we can see, DNS-SD uses multiple TXT entries, while Link Format uses only one, allowing for a reduction in overload for the transmission of each record, due to the restrictions in the payload size for the Internet of Things communication technologies.

An example of this access via CoAP is the following request for the resource type Temperature over a determined resource directory, such as:

GET /rd.esiot.com?rt=Temperature

The result could indicate the CoAP method "temp" over the URL node1.esiot.com for the asked resource type Temperature.

<coap://node1.esiot.com/temp>;rt="Temperature"

In this particular case, native access from DNS will be understood to request the records. Figure 1 shows the example of a device connected by way of a native interface "if" IEEE 802.15.4, with the CoAP methods "value" to consult the current status of the sensor, and the method "onoff" to change the status, as well as common fields defined by the CoRE Link Local format, such as the resource type "rt", and instance "ins".

In addition to this basic approach, in this work we have defined an extended set of parameters for the "TXT" record. We should highlight that it has a capacity of up to 65536 bytes. For example, Fig. 8 presents the look-up for a light under rd.esiot.com. In particular, the light2.rd.esiot.com, which is a resource type of dimmer with EIB/KNX technology, and the location for the specific technology, which is accessed through the platform (presented in Section 5) is Area: 1, Zone: 2 and DeviceID: 3. Therefore, this offers an extended description of the sensor and its status. All of these values can be consulted through CoAP methods, as defined with the CoRE approach as well.

Furthermore, it should be noted that DNS has some limitations regarding automatic updates, and dynamic and fluid interactions with devices. In order to stay synchronized with the current status of the sensor, the management of the domain has been integrated inside the AAID gateway, that is, the manager for the mapping between the IPv6 address and the specific device. Automatic updates and synchronization with the DNS values are performed, together with the real status from the network. As presented in Section 5.2, these functions have been specifically implemented in the Border Router from the Internet of Things network, which is the same gateway where the AAID gateway is located.

It can be located outside the AAID gateway/Border Router as well, but this could cause the above-mentioned synchronization problems. If DNS-SD is not located at the gateway, it can define multiple ways to discover this resource directory location, such as using DHCP.

Finally, the example for the light1.rd.esiot.com, and light2.rd.esiot.com with EIB/KNX and X10 technology (see Fig. 8) has been presented; this platform offers connectivity with different non-IP technologies, such as EIB/KNX, X10, Bluetooth, IEEE 802.15.4, etc. This allows for managing the information locally, even when it is globally available and accessible through DNS queries.

## 5. Evaluation

The AAID Gateway has been implemented in a multi-protocol system based on an ARM CPU and an embedded Linux OS, and the AAID node in an IEEE 802.15.4 USB dongle based on Jennic JN5139 chip. Furthermore, the performance and scalability of this system has been evaluated, together with the interactivity and mapping process between the AAID node and the AAID Gateway over the IPv6 network from the University of Murcia.

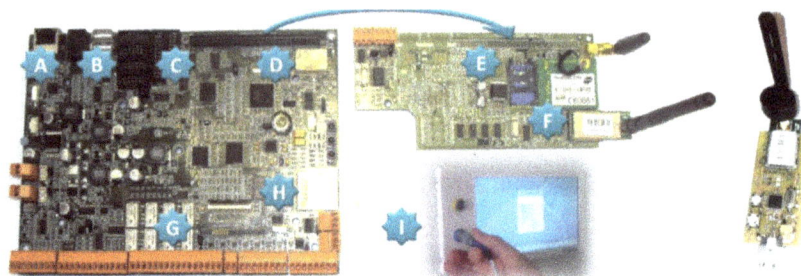

Fig. 9. AAID Gateway on the multiprotocol card (left) and AAID node on the 802.15.4 USB dongle (right).

## 5.1. Implementing Glowbal IP

The evaluation was conducted on the platform that our research lab developed to build automation and industrial environments. For example, a previous version of this platform was deployed in a building automation solution [41],

This platform consists of a hardware modular solution, based on Linux OS, that allows the upgrade of platform components without requiring reconfiguration of the other components. This characteristic makes this platform more expandable, and offers the option of installing software that has already been tested and made available for Linux, making it more robust. Specifically, it supports capabilities for routing (*route*), router advertisement (*radvd*), policies (*iptables*), and DNS-SD (*bind*).

The implementation of the daemon involved in the AAID gateway makes this adaptation practicable. Specifically, a *tap* interface is built with the addresses assigned to the objects with AAID. Thereby, it is able to collect all the received packets for AAID nodes through a daemon with a RAW socket, *SOCK_RAW option*, listening in promiscuous mode, *ETH_P_ALL option*, in the *tap* interface.

The AAID Gateway platform is presented on the left side of Fig. 9. It is based on the ARM9@400Mhz 32-bit processor, with 256MB LPDDR RAM memory, and 256MB NAND memory, which supports Linux OS. It offers an Ethernet 10/100Mbps (A), 2 USB 2.0 ports (B), 4 Serial RS232 ports (C), Bluetooth 2.1 with HDP profile compliant with BlueGiga (D), GPRS from WaveCom (E), IEEE802.15.4 ZigBee from Jennic (F), 24 inputs/outputs among digitals/analogs/relays (G), a compact flash support for data logging (H), and a touch screen LCD (I). Finally, other capabilities are interesting for continuous sensing, such as real-time watch, 5 high precision timers, 2 analog/digital converters for analog signal processing, a random number generator for security seeds, and IPv6 stack support.

Furthermore, the AAID node is shown on the right side of Fig. 9. It is based on the Jennic JN5139 module. It also has a OpenRISC 32-bit processor, which supports IEEE802.15.4 stack, and ZigBee Pro. It is able to set it up with 6LoWPAN as well, but for this evaluation it has been defined with the basic IEEE 802.15.4 stack. This is connected to the PC through the USB port. Thereby, it is easily programmable and permits debugging the results through the emulated serial port. In addition, it offers an advanced cryptography stack based on Elliptic Curve [42], which is highly relevant for the Internet of Things, due to its security requirements and privacy issues [43,44].

The next section shows the evaluation performed over this platform, where the AAID approach has been implemented.

## 5.2. AAID mapping performance

The AAID mapping performance has been evaluated in the multi-protocol card shown in Fig. 9 (left). This multi-protocol board is responsible for mapping between the multiple technologies and its corresponding IPv6 addresses, and vice-versa.

This AAID Gateway functionality is divided into the following two modules:

– *Sender module*: This corresponds to the module allocated in the multiprotocol card to manage the traffic generated from the non-IP devices, by way of AAID technology support to the Core Network. This module identifies and knows in-depth the connection and session information for each device, since it manages the L2G table. It is listening to the native interfaces from the technology used for each device – IEEE 802.15.4 in this case, through a serial port – but it can be also accessed through a serial port in board technologies such as X10, EIB, CAN, and Bluetooth. It checks that the AAID is previously used and available in the L2G, or if it is a set message and requires inclusion in the L2G mapping table.
– *Receiver module*: This is also allocated in the multiprotocol card to manage the incoming traffic from the Core Network to the specific technologies and specifications. This module is a daemon which is listening, in promiscuous mode, to the messages received from the global network. Once a message is received, this receiver maps the IPv6 destination address to the correspondent AAID, and follows the mapping specifications in order to detect whether this is from a previous session, or whether it is necessary to set up a new AAID.

In the AAID node, a pre-analysis in the incoming packets is also defined to determine the session from the coming packet, as well as a post-processing, in order to include the AAID information to the payload.

This process has been evaluated with a flow of data packets between a server and an AAID node with 500 consecutive requests (each one with a different port). Over this evaluation, the total time was calculated for processing the mapping and the reply. Therefore, the transmission time has been calculated as the difference between the total time, and the time calculated locally in the AAID GW, and AAID for the processing. The result is divided by 2, to consider only the time for one-way transmission, not for the full round-trip.

The results, shown in Fig. 10, come from the division in processing in the AAID GW, the processing in the AAID node, and the one way transmission time (half of the round-trip transmission time).

Furthermore, a summary of the results has been also defined in Table 2, where the minimum, maximum, and average for each of the considered times were calculated.

With regard to the mapping processing, it is not introducing a high delay. More specifically, the mapping processing in the AAID gateway is practically instantaneous; being only slightly higher in the AAID node due to its constraints, but this also presents an average time of 22 milliseconds. Therefore, carrying out this mapping processing is entirely suitable.

It is important to point out that the transmission times are low, since they were evaluated with a server located at the same network. But, it is not relevant for the evaluation for the AAID mapping.

## 6. Discussion

This section summarizes how Glowbal IP addresses the requirements from Internet of Things to extend the access and use of the Future Internet infrastructure, homogenizes the access to the services/resources/devices, and extends global access with high level aspects such as interoperability among scenarios, as well as future mobility and multi-homing. Specifically, it can be summarized in five key points, which represent a significant contribution to the current status of the field.

At the beginning, Glowbal IP followed the spirit from Future Internet; IPv6 nowadays is being turned into a key technology for the Internet of Things. Until now, the usual practice was build islands of Wireless Sensor Networks connected to the Internet by way of a Portal Server or a Gateway, where the

Table 2
Summary of the mapping performance for 500 sessions

| Description | Minimum | Maximum | Average |
|---|---|---|---|
| Time transmission (half of a round-trip) | 9,900 ms | 32,920 ms | 15,900 ms |
| Time processing AAID Gateway | 0,020 ms | 0,110 ms | 0,033 ms |
| Time processing AAID node | 20,828 ms | 34,170 ms | 22,828 ms |
| Total | 22,444 ms | 54,965 ms | 38,969 ms |

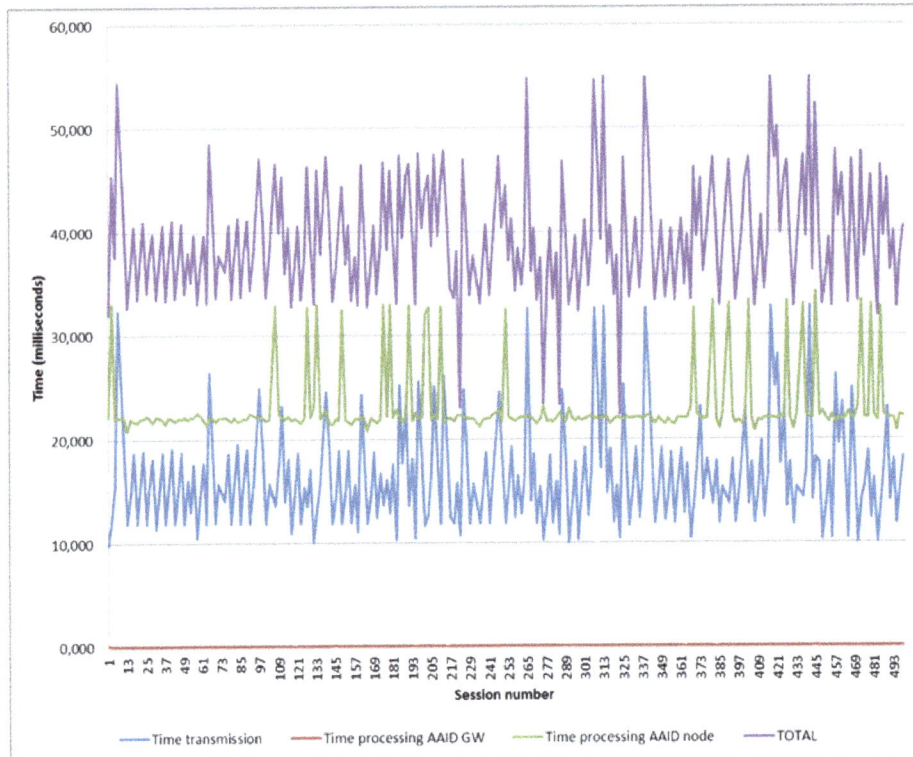

Fig. 10. Performance evaluation for the AAID mapping in AAID Gateway and AAID node.

information is processed and later sent to the defined clients. These approaches offer what literature calls an Intranet of Things. Therefore, in order to move toward a real Internet of Things, it is necessary to provide support for end-to-end and global access, which is located at IPv6 capabilities.

The second contribution is opening legacy technologies and already deployed devices to a new range of IPv6-enabled services. Based on the Glowbal IP integration mentioned in the first contribution, a real interconnected physical and virtual object (end-to-end) can be reached by way of IPv6. Thereby, a homogeneous, transparent and scalable access to the devices and services can be achieved. From the Web of Things approach, the RESTFul methods could be applied directly to the end-device, increasing the scalability of the solution, flexibility and allowing for extension of the ubiquitous concept with mobility and global interoperability.

The third contribution is that Glowbal IP supports RESTFul/CoAP at the end-device level, in order to make the solution more scalable than the current approach mentioned, where the portal server is required to translate from native protocol (e.g. Waspmote) to HTTP Push method (e.g. REST). In short, Glowbal IP presents the integration of CoAP methods in the IoT nodes.

The fourth contribution is the interoperability of heterogeneous Smart Things (IoT nodes), systems and scenarios from different IPv6 enabling technologies, such as 6LoWPAN, and native IPv6. The mentioned enablers are supplemented with Glowbal IP for other nodes such as IEEE 802.15.4 and Bluetooth nodes. The interoperability is reached through homogenous RESTFul/CoAP. In addition to the interoperability among scenarios, communication with external sensors is also feasible. Therefore, it also enables cross-domain applications. It is well known that IoT technology is useful in numerous application domains, ranging from environmental monitoring to health monitoring, smart spaces, improvement of industrial processes, and so on. The entry barrier to getting involved in this domain and starting to offer new products or services is not very high: when the plethora of potential applications and low entry barrier is combined, the IoT domain is suitable for agile SMEs to develop a range of new products and services. An easy integration of the current SME solutions with the services and potential offered by IoT is also feasible. All in all, it is opening possibilities to define innovative business models based on end-to-end connectivity and supporting innovative Internet services.

The fifth contribution is the re-use of existing IP technologies such as DNS Service Directory (DNS-SD) and multicast DNS (mDNS) for discovery purposes, in addition to defining the exploitation of network-based information systems through the definition of the resource and service directories in the already deployed DNS servers.

## 7. Conclusions and future works

This paper has presented the innovative protocol Glowbal IP for the extension of devices already deployed and legacy technologies to IPv6. It allows a user to create global communications and network layer homogeneity through IPv6. In addition, Glowbal IP solves the lack of optimization from 6LoWPAN with regard to global communications, which requires from 26 to 41 bytes for the 6LoWPAN header, while AAID only requires a 5-byte header, when the identification of the session is carried out by way of AAID identifiers.

The research has also shown how to exploit the network-based information systems through the current DNS deployment, with the DNS Service Directory and multicast DNS. This allows a system to describe information in the network about devices such as family, model, features and Web Services offered in order to interact with it.

Finally, this has been evaluated in a multiprotocol card developed in our lab, and this system represents a low delay because the mapping tasks defined Glowbal IP as a suitable technology to be deployed.

Ongoing work is focused, firstly, on the extension of Glowbal IP for mobility and multi-homing, and the connection of islands of discoverability. For these three purposes, we are working on an overlay which offers ancillary support to the capabilities from IPv6 protocols. Secondly, Glowbal IP will be evaluated for Bluetooth Low Energy, due to its capabilities for the Internet of Things. Specifically, this is an interesting technology because it is located in handset devices, and these devices offer connectivity with the Internet through their GPRS/GSM network interfaces. Therefore, they can be used as a gateway for the management of a personal network.

## Acknowledgments

This work was made possible by the generous support of the Excellence Research Group Program (04552/GERM/06), from Seneca Foundation; FPU program (AP2009-3981), from the Spanish Ministry of Education and Science; and in the framework of the IoT6 European Project (STREP) from the 7th Framework Program (Grant 288445).

# References

[1]   C. Endres, A. Butz and A. MacWilliams, A survey of software infrastructures and frameworks for ubiquitous computing, *Mobile Information Systems* **1**(1) (2005), 41–80.
[2]   L. Atzori, A. Iera and G. Morabito, The Internet of Things: A survey, *Computer Networks* **54**(15) (2010), 2787–2805.
[3]   A. Durresi, M. Durresi, A. Merkoci and L. Barolli, Networked biomedical system for ubiquitous health monitoring, *Mobile Information Systems* **4**(3) (2008), 211–218.
[4]   C. Sotomayor, A.J. Jara and A.F.G. Gómez-Skarmeta, Real-Time Monitoring System for Watercourse Improvement and Flood Forecast, Lecture Notes in Computer Science, *Springer Verlag* **6935** (2011), 311–319. ISSN: 0302-9743.
[5]   J. Santa, M.A. Zamora, Antonio J. Jara and A.F.G. Skarmeta, Telematic platform for integral management of agricultural/perishable goods in terrestrial logistics, *Computers and Electronics in Agriculture*, ElSevier, ISSN: 0168-1699, Vol, 89, pp. 31–40. doi:10.1016/j.compag.2011.10.010, 2011.
[6]   O. Haubensak, Smart Cities and Internet of Things, Business Aspects of the Internet of Things, 2011.
[7]   A.J. Jara, M.A. Zamora and A.F.G. Skarmeta, An internet of things-based personal device for diabetes therapy management in AAL, *Personal and Ubiquitous Computing* **15**(4) (2011), 431–440.
[8]   Z. Shelby, Embedded web services, Wireless Communications, *IEEE* **17**(6) (December 2010), 52–57, doi: 10.1109/MWC.2010.5675778.
[9]   G. Montenegro, N. Kushalnagar, J. Hui and D. Culler, Transmission of IPv6 Packets over IEEE 802.15.4 Networks, RFC4944, September 2007.
[10]  J. Hui and P. Thubert, Compression Format for IPv6 Datagrams over IEEE 802.15.4-Based Network, IETF 6LoWPAN Working Group, RFC6282, 2011.
[11]  B. Emerson, M2M: the internet of 50 billion devices, Win-Win, Editorial: Huawei, January 2010.
[12]  J.J.P.C. Rodrigues and P.A.C.S. Neves, A Survey on IP-based Wireless Sensor Networks Solutions, in: *International Journal of Communication Systems*, Wiley, ISSN: 1074-5351, Vol. 23, No. 8, August 2010, pp. 963–981.
[13]  IEEE Standard for Information technology, Telecommunications and information exchange between systems – Local and metropolitan area networks – Specific requirements. Part 15.4: Wireless Medium Access Control (MAC) and Physical Layer (PHY) Specifications for Low-Rate Wireless Personal Area Networks (WPANs), IEEE 802.15.4, 2006.
[14]  M. Goyal, W. Xie and H. Hosseini, IEEE 802.15.4 modifications and their impact, *Mobile Information Systems* **7**(1) (2011), 69–92, 10.3233/MIS-2011-0111.
[15]  A.J. Jara, L. Marin, M.A. Zamora and A.F.G. Skarmeta, Evaluation of 6LoWPAN capabilities for secure integration of sensors for continuous vital monitoring, V International Symposium on Ubiquitous Computing and Ambient Intelligence (UCAmI'11), 2011.
[16]  H.K. Kahng, D.-I. Choi and S. Kim, Global connectivity in 6LoWPAN, draft-kahng-6lowpan-global-connectivity-01.txt, IETF work in progress, March, 2011.
[17]  B. Gohel and D. Singh, Global connectivity for 6lowpan draft-singh-6lowpan-global-connectivity-01.txt, IETF work in progress, Feb 2011.
[18]  W.K. Edwards, Discovery systems in ubiquitous computing, Pervasive Computing, *IEEE* **5**(2), 70–77, doi: 10.1109/MPRV.2006.28, 2006.
[19]  S. Kiyomoto and K.M. Martin, Model for a Common Notion of Privacy Leakage on Public Database, *Journal of Wireless Mobile Networks, Ubiquitous Computing, and Dependable Applications* (*JoWUA*) **2**(1) (2011), 50–62.
[20]  A.J. Jara, R.M. Silva, J.S. Silva, M.A. Zamora and A.F.G. Skarmeta, Mobile IP-based Protocol for Wireless Personal Area Networks in Critical Environment, Wireless Personal Communications, Springer London, ISSN: 0929-6212, Vol. 61, No. 4, 2011, pp. 711–737.
[21]  A. Herstad, E. Nersveen, H. Samset, A. Storsveen, S. Svaet and K.E. Husa, Connected objects: Building a service platform for M2M, Intelligence in Next Generation Networks, 2009. ICIN 2009. 13th International Conference on, doi: 10.1109/ICIN.2009.5357057M2M platforms, 2009.
[22]  J. Hodges and R. Morgan, Lightweight Directory Access Protocol (v3), IETF RFC 3377, Sept. 2002.
[23]  P. Mockapetris, Domain Names – Concepts and Facilities, IETF RFC 1034, Nov. 1987.
[24]  OASIS, Introduction to UDDI: Important Features and Functional Concepts, Organization for the Advancement of Structured Information. Standards (OASIS), Oct. 2004.
[25]  S. Cheshire and M. Krochmal, DNS-Based Service Discovery, IETF Zeroconf Working Group, www.zeroconf.org/ and www.dns-sd.org/, draft-cheshire-dnsext-dns-sd.txt, 2011.
[26]  T.R. Sheltami, E.M. Shakshuki and H.T. Mouftah, A web-based application of TELOSB sensor network, *Mobile Information Systems* **7**(2) (2011), 147–163, 10.3233/MIS-2011-0115.
[27]  Z. Shelby, K. Hartke, C. Bormann and B. Frank, Constrained Application Protocol (CoAP), draft-ietf-core-coap-06, IETF work in progress, May 2011.
[28]  Z. Shelby and S. Krco, CoRE Resource Directory, draft-shelby-core-resource-directory-02, IETF work in progress, 2011.

[29]  M. Sabou, Smart objects: Challenges for Semantic Web research, *Semantic Web* **1**(1) (2010), 127–130, doi: 10.3233/SW-2010-0011.

[30]  Z. Shelby, CoRE Link Format, draft-ietf-core-link-format-06, IETF work in progress, June 2011.

[31]  A.J. Jara, P. López Martínez, D. Fernández Ros, B. Úbeda, M.A. Zamora and A.F.G. Skarmeta, Heart monitoring system based on NFC for continuous analysis and pre-processing of wireless vital signs, *International Conference on Health Informatics*, HEALTHINF, 2012.

[32]  A. Durresi, P. Zhang, M. Durresi and L. Barolli, Architecture for mobile Heterogeneous Multi Domain networks, *Mobile Information Systems* **6**(1) (2010), 49–63, 10.3233/MIS-2010-0092.

[33]  L.M.L. Oliveira, A.F. de Sousa and J.J.P.C. Rodrigues, Routing and Mobility Approaches in IPv6 over LoWPAN Mesh Networks, in: *International Journal of Communication Systems*, Wiley, ISSN: 1074-5351, Vol. 24, Issue 11, November 2011, pp. 1445–1466.

[34]  F.-Y. Leu, I. You and F. Tang, Emerging Wireless and Mobile Technologies, *Mobile Information Systems* **7**(3) (2011), 165–167, 10.3233/MIS-2011-0122.

[35]  W. Wu, X. Li, S. Xiang, H.B. Lim and K.-L. Tan, Sensor relocation for emergent data acquisition in sparse mobile sensor networks, *Mobile Information Systems* **6**(2) (2010), 155–176, 10.3233/MIS-2010-0097.

[36]  T. Narten and T. Jinmei, IPv6 Stateless Address Autoconfiguration, IETF Network Working Group, RFC 4862, 2007.

[37]  S. Deering, IP Version 6 Addressing Architecture, IETF Network Working Group, RFC 4291, 2006.

[38]  M. Crawford, Transmission of IPv6 Packets over Ethernet Networks, IETF Network Working Group, RFC 2464, 1998.

[39]  D. Guinard, V. Trifa, S. Karnouskos, P. Spiess and D. Savio, Interacting with the SOA-Based Internet of Things: Discovery, Query, Selection, and On-Demand Provisioning of Web Services, Services Computing, *IEEE Transactions on* **3**(3) (2010), 223–235, doi: 10.1109/TSC.2010.3.

[40]  A. Gulbrandsen, P. Vixie and L. Esibov, A DNS RR for specifying the location of services (DNS SRV), IETF RFC 2782, http://www.dns-sd.org/ServiceTypes.html, 2000.

[41]  M.A. Zamora, J. Santa and A.F.G. Skarmeta, An integral and networked Home Automation solution for indoor Ambient Intelligence, *IEEE Pervasive Computing* **9** (2010), 66–77.

[42]  L. Marin, A. Jara and A. Skarmeta, Shifting Primes: Extension of pseudo-Mersenne primes to optimize ECC for MSP430-based Future IoT devices, Multidisciplinary Research and Practice for Business, Enterprise and Health Information Systems, Springer, LNCS, 2011.

[43]  T.A. Zia and A.Y. Zomaya, A Lightweight Security Framework for Wireless Sensor Networks, *Journal of Wireless Mobile Networks, Ubiquitous Computing, and Dependable Applications* (*JoWUA*) **2**(3) (2011), 53–73.

[44]  R. Roman, P. Najera and J. Lopez, Securing the Internet of Things, *Computer* **44**(9) (Sept 2011), 51–58.

**Antonio J. Jara-Valera** received his B.S. (Hons. – valedictorian) degree in Computer Science from the University of Murcia (UMU), Murcia (Spain) in 2007, and his M.S. degree in Computer Science from the same institution in 2009, where his Master thesis was about "Internet of Things in clinical environments". He received a second M.S. degree in Computer Science from the University of Murcia in 2010, focused on advanced networks and artificial intelligence, and whose Master thesis was about "Mobility protocols for 6LoWPAN". He has collaborated with UMU's Department of Information Technology and Communication Engineering since 2007, where he currently is working on several projects related to the ZigBee/6LoWPAN and RFID applications in Intelligent Transport Systems (ITS), home automation and mainly healthcare. He is especially focused on IPv6 integration, security and mobility for Future Internet and Internet of Things, which are the topics of his Ph.D. He has published over 40 international papers in this area, and he has worked on different research projects in the national and international area for Internet of Things and IPv6 integration in e-Health, ITS, and building automation.

**Miguel A. Zamora-Izquierdo** received the M.S. degree in automation and electronics and the Ph.D. degree in computer science from the University of Murcia, Spain, in 1997 and 2003, respectively. Since 1999, he has been an Associate Professor with the Department of Information and Communication Engineering, UMU, where he works on several projects related to the remote monitoring and control with a focus on sensors system and embedded system.

**Antonio Skarmeta** received the M.S. degree in Computer Science from the University of Granada and B.S. (Hons.) and Ph.D. in Computer Science from the University of Murcia, Spain. Since 2009 he is full professor at the same department and University. Antonio F. Gómez-Skarmeta has worked on different research projects in the national and international area, like Euro6IX, 6Power, Positif, Seinit, Deserec, Enable, Daidalos, ITSS6, and IoT6. He is mainly interested in the integration of security services at different layers like networking, management and web services. Associate editor of the IEEE SMC-Part B and reviewer of several international journals, he has published over 90 international papers and is member of several program committees.

# 9

# Smart object reminders with RFID and mobile technologies

Hui-Huang Hsu[a,*], Cheng-Ning Lee[a], Jason C. Hung[b] and Timothy K. Shih[c]

[a]*Department of Computer Science and Information Engineering, Tamkang University, 151 Ying-Chuan Rd., Tamsui, New Taipei City, Taiwan*

[b]*Department of Information Management, Overseas Chinese University, 100 Chiao Kwang Rd., Taichung, Taiwan*

[c]*Department of Computer Science and Information Engineering, National Central University, 300 Jhongda Rd., Jhongli, Taoyuan, Taiwan*

**Abstract.** In this paper, we present a reminder system that sends a reminder list to the user's mobile device based on the history data collected from the same user and the events in the user's calendar on that day. The system provides an individualized service. The list is to remind the user with objects he/she might have forgotten at home. The objects that the user brings along with are detected by passive RFID technology. Objects are classified into three different levels based on their frequencies in the history data. Rules of the three levels are then followed to decide if a certain object should be in the reminder list or not. A feedback mechanism is also designed to lower the possibility of unnecessary reminding.

Keywords: RFID, reminder systems, personalization, smart home, ambient intelligence

## 1. Introduction

Sensors are the basic components needed to detect the contextual needs of the users. In recent years, sensor network research has been a hot topic [1–3]. In a smart home, sensors are deployed at home to collect user-centered information in the environment. The collected information is then processed by the server to provide suitable services to the user. This is also a part of *ambient intelligence*. Some people tend to forget things when they leave home for work or school. It is desirable that people are reminded with the things they forgot to bring along with. In this paper, we propose a smart reminding mechanism using RFID (Radio Frequency Identification) and mobile technologies. The mechanism is designed to fit individual needs from the history data and the event calendar of the user.

Previously, RFID was used for abnormal behavior detection in elderly care [4] and object reminders to school kids [5]. According to the survey of IDTechEx, the applications of RFID will reach the highest demand from 2015 to 2020. Although the current RFID software market is not so big, with the growing popularity of RFID the RFID software market will have a potential high growth in the near future [6]. Research in location-based queries in a mobile environment can be found in [7,8]. For further researches on mobile services, the readers are referred to [9–11].

---

*Corresponding author. E-mail: h hsu@mail.tku.edu.tw.

In smart home research, Future Home is a real environment built by Microsoft [12]. It includes the front door, the entry/foyer, the kitchen, the family room, the dining room, the entertainment room and the bedroom. In Japan, Toyota also built a future home called PAPI [13]. The research combined automatic cars and IT facilities produced by Toyota. In Taiwan, Farglory Realty used advanced IT technologies for home security, such as face recognition and finger vein identification.

This research aims at providing a reminding mechanism for the user in a smart home environment. We focus on objects that the user would bring along when he/she goes out. The RFID technology is used to sense the objects the user brings along at the front door. In the object database, an object is recorded not only by its name along with a unique RFID number, but also its class. The taxonomy has two levels: classes and objects. For example, both the June issue and the July issue of Scientific American have the same class – Scientific American. Furthermore, both classes and objects are classified into three different levels according to the frequencies the user took them out. The three levels are *daily*, *weekly* and *non-regular*. The reminder system is also connected to the user's calendar. Objects related to events on the calendar are also recorded in the database. An event object list can be built by the user or automatically detected and recorded by the system.

When the user leaves home, the reminder system checks the objects in his/her bags and pockets, and compares the objects with a list of objects generated by the system according to the day of the week and the events on the calendar. The system then sends a reminder object list to the mobile phone or PDA (Personal Digital Assistant) of the user. The user can quickly browse the list and take any objects he/she might have forgotten to bring along. There is also a feedback mechanism for the reminder list. If any objects mistakenly appear on the list, the user can give a feedback to the system. The user can simply mark the unneeded objects on the reminder list. Those objects will not be included in the reminder list in the same situation next time. In this research, we have implemented a prototype system for the ideas. The preliminary tests show that this approach is promising for real applications.

In Section 2, related work in RFID applications and smart environment is discussed. The proposed smart reminding mechanism is introduced in Section 3. System implementation and tests are then presented in Section 4. Finally, a brief conclusion is drawn in Section 5.

## 2. Related work

### 2.1. RFID applications

There are three major components of an RFID system: the reader, the antenna, and the tags. Each tag is associated with a unique number. When a tag is in the detection range of the reader, the number is read. Two types of tags can be found: active tags with a longer detection range and passive tags with a shorter detection range [14]. An RFID tag is usually attached to an object and the information of the object along with the RFID number are recorded in the database. Whenever the RFID tag is sensed, the object can thus be identified.

The papers to be discussed here are more advanced applications. One of the papers deploys a large amount of RFID tags in an office, a conference venue or other public places. The user can use mobile devices to receive desired information like locations or maps, also the user can leave a message to a certain person. Besides, the maps can be retrieved by the rescue crew in emergency through a tag [15]. RFID was also applied to home cooking. First, RFID tags are attached to ingredients and utensils. The system can automatically provide video instructions to the user according to the detected movements of the user. The user needs not check the cook book step by step [16].

Another paper proposed a client-server architecture that can remotely control home appliances via mobile devices. Massive RFID tags are distributed in the environment for location awareness. The advantages of this system are less power consumption and design complexity [17]. In medical care, there is an RFID application related to medicine taking. RFID tags are attached to medicine bottles and the reader is placed in the drawer storing the medicine. The system can help the elderly people to record their medicine-taking data and determine whether they have taken the right medication with right dosage [18]. On the other hand, the assessment of independent living ability for the elderly can also be done by the RFID technology. RFID tags are attached to the tools used in our daily life. Readers are placed on different body parts of the elder. Home activities of the elder can thus be recorded [19,20]. Above-mentioned researches were keen to use the RFID technology to provide the users with convenient services. Although the RFID tags deployed in the public areas can help reach the goal, the tag price is still too high to make it feasible in real applications. In addition, using RFID to detect user's activities requires the user to carry RFID readers. This is very inconvenient to the user. Hopefully, the RFID reader can become much smaller and light-weighted in the near future.

## 2.2. Smart environment

According to Mark Weiser, there would be many invisible sensors, actuators, computers and displays, embedded into our daily lives and contacted each other via the Internet in the real world. Smart environment is defined to be a small world with numerous smart devices constantly working to provide comfort to the inhabitants [21–23]. This concept is commonly used in residential. Researchers hope to use machine learning methods to learn the living habits of each inhabitant, and intelligent agents to provide services to each user. Generally speaking, these studies have to collect enough information/data first to train the machine learning system. For example, inhabitant behaviors at home can be recorded with a time stamp. The relationship of the behaviors in time can thus be analyzed [24–26]. Another issue is that people's behavior and demand usually change with time. Therefore, the system must also adapt to each user. A research used cameras, microphones and other wearable sensors to track the inhabitant. The support vector machine (SVM) was then used to determine the user's behavior change and hence different environment settings [27]. A good system not only can changes with the user, but also can predict the next user activity. Another research focused on patients with dementia and loss of short-term memory. A system was designed to remind the patients if a certain action has been taken (such as having tea, brushing teeth, and… etc.). When the patient forgets the next action, the system can give tips through LED lights, images or text to remind the patient [28].

## 3. Smart reminding mechanism

A smart reminding mechanism is designed to remind the user what he or she has forgotten to bring along when leaving home for work or school. The user can thus takes the needed objects before going far from home. The RFID technology is used here to sense the objects. There are two ways to determine the list of objects a person needs for daily work. First, if the events on that day are known, objects for those events should be in the list. Secondly, if the objects that the user used to bring along with on that certain day of the week, they should also be in the list. In this research, we tried to combine these two sources of information to construct the reminder list for the user when he/she goes out for work or school on a certain day.

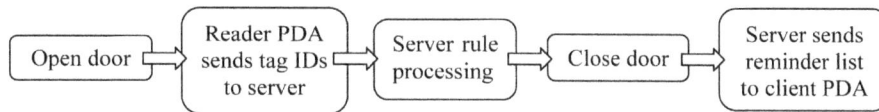

Fig. 1. Workflow of the reminder system.

The background setting of the developed system is in a general home environment. An RFID reader is installed near the front door and RFID tags are attached to the objects the user would take out. Whenever the user goes out, the objects he/she brings along with are detected by the reader and saved in the database. In general, the objects are strongly related to the day in a week. For example, Jack is a student and he has Calculus classes on Wednesday. Therefore, he usually brings the Calculus textbook with him on Wednesday. Max plays tennis with his friends on Friday evening. He brings his tennis racquet with him on Friday.

## 3.1. Reminder server

Figure 1 illustrates the workflow of the reminder system. When the front door is open, the RFID reader is triggered to read the nearby tags and send their ID numbers to the server. The maximum detection distance should be set at around 1 meter. The server compares the list with the object list generated by a number of rules following the historical data and the calendar events. The missing objects are arranged in a reminder list. When the door is closed, the server sends the list to the mobile phone or PDA of the user. The functions of the reminder server and the PDA client are discussed in the following.

(a) Collect tag IDs from the RFID reader.
(b) Process the rule sets.
(c) Generate and send the reminder list to the PDA client.

The user needs to initiate the system by inputting the object data including the object name, its RFID tag ID, and its class to the object database in the server. Recall that the objects are placed in a two level taxonomy: classes and objects. When the server received the tag IDs from the reader (detected when the user leaves home), it records the date along with the tag IDs. The tag IDs are associated with the objects and their classes in the database. The object/class use records for a week are then processed by the rule sets discussed in the following.

There are three sets of rules: the object rules, the class rules and the event rules. The object rules are used to classify all objects into one of the three frequency levels: *daily*, *weekly* and *non-regular* (Table 1). *Daily* means that the object is taken out by the user every day, e.g., keys and wallet. *Weekly* means that an object is needed on certain days during a week, e.g., English textbook on Monday and Wednesday because there are English classes on the two days. The days can be found from the use records in the previous week and they are recorded with the object in the database. As for *non-regular*, the objects have not been detected in the past seven days. The rules also define the conditions for object level change. An object can be upgraded or downgraded in the three levels. The adjustment periods of upgrade and downgrade can be made daily or weekly. For daily checks, the system makes adjustments right after the user leaves home every day. For example, if a non-regular object is brought by the user today, it is upgraded to a weekly object right away. If a weekly object is not brought by the user today, it is downgraded to a non-regular object. For weekly checks, the rules are performed every seven days. For example, if a weekly object was taken every day in the past week, it is upgraded to a daily object. If a daily object was not taken in any days in the past week, it is downgraded to a weekly object.

Table 1
Rules for object classification, upgrade and downgrade

| Object classification | Classification condition | Upgrade condition | Downgrade condition |
| --- | --- | --- | --- |
| Daily | Object is brought every day in the past week. | Not applicable | Object is not brought for some days in the past week. (Check every week) |
| Weekly | Object is brought 1 to 6 days in the past week. | Object is brought every day in the past week. (Check every week) | Object is not brought in 7 consecutive days. (Check every day) |
| Non-regular | Object is not brought in the past 7 days. | Object is brought today. (Check every day) | Not applicable |

The second rule set is called class rules. A class is a collection of objects with similar features. For example, different issues of the same magazine are of the same class. Classes are also classified into daily, weekly and non-regular and the object classification rules in Table 1 are applicable to the classes. Therefore, the reminder list generated by the server includes not only the objects, but also the classes. For example, if a user takes Business Week with him/her every day, it is certain that the system will remind him/her whenever he/she forgets to bring *any* issue of Business Week on that day. Moreover, for the same medicine, every bottle has its unique tag ID. The user might need the medicine every day, but he/she might bring different bottles. If all the bottles are set to the same medicine class, the system can remind the user by the medicine class, not a particular bottle. The upgrade and downgrade rules in Table 1 also apply to the classes, but they are processed only in a weekly basis.

The third rule set is event rules. The rules help record the objects a user needed for a certain event. The objects for an event can be determined by the historical data and adjusted by the user through the feedback mechanism or they can be input directly by the user. When there is an event today and the event objects do not exist in the database. The event objects can be determined by subtracting daily and weekly objects from the objects that the user brings with him/her today. The server will automatically add the objects to the event. This can be illustrated by the following equation: *Event_Objects = All_Objects − Daily_Objects − Weekly_Objects*. If there are more than one unknown events, this might cause incorrect event object lists. However, they can be corrected with the user feedback at a later time. This process is shown in the upper half of the flowchart in Fig. 2. The object rules and the class rules are performed in the lower half of the flowchart.

Finally, a reminder list is sent to the user's PDA or smart phone before he or she goes far. The objects in the reminder list are ordered as follows: 1) event objects, 2) daily objects, 3) weekly objects, 4) daily classes, and 5) weekly classes. Meanwhile, level upgrade and downgrade rules are applied and related records are updated.

### 3.2. PDA client

A PDA is used in our research as the client. A smart phone is another possibility. There are three main functions of the client.

(a) Reminder: When the user leaves home, the system automatically compares the taken objects with the object list generated by the system. The system then sends the reminder object/class list to the user's PDA. The PDA client is responsible for displaying the list to the user.

(b) Calendar: The user can manage events in his/her personal calendar on the PDA. The system will automatically update the calendar of the user on the server.

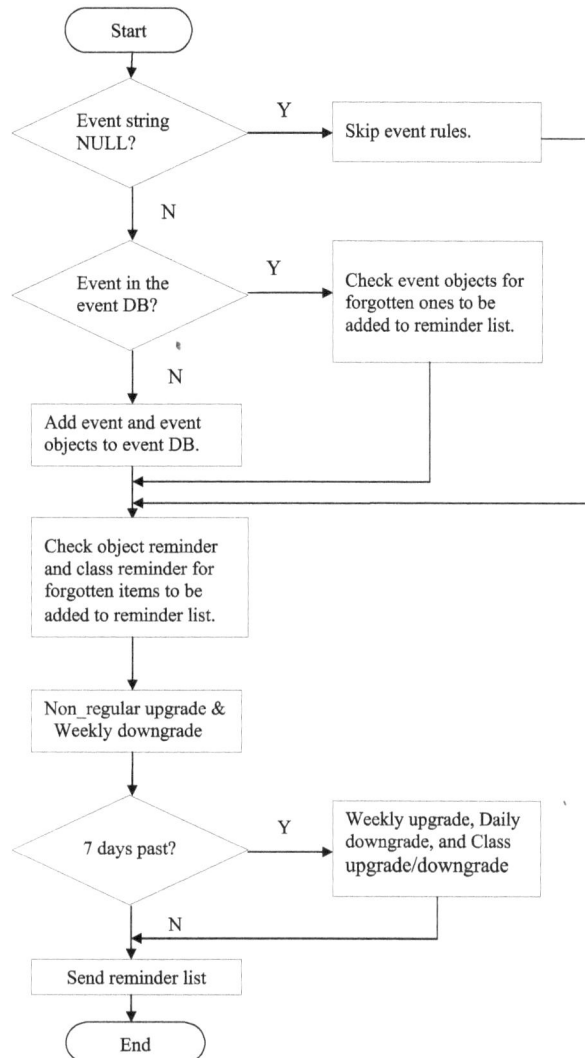

Fig. 2. The flow chart of rule processing.

(c) User feedback: If the user is not satisfied with the reminder list, he/she can decide whether to ignore some of the objects in the reminder list and give a feedback to the system. The object/class frequency levels are then adjusted accordingly. Also, the user can modify the event objects for each event. With this feedback mechanism, the reminder system can provide a better service to the user with a more accurate reminder list.

## 4. System implementation and tests

Passive tags are first attached to the objects in our implementation. They are used, rather than active tags, because detection range is more appropriate for this application. Also, passive tags are smaller and lighter than active tags. Figure 3 shows a few examples. The RFID tags follow the ISO15693

Fig. 3. Objects attached with an RFID tag.

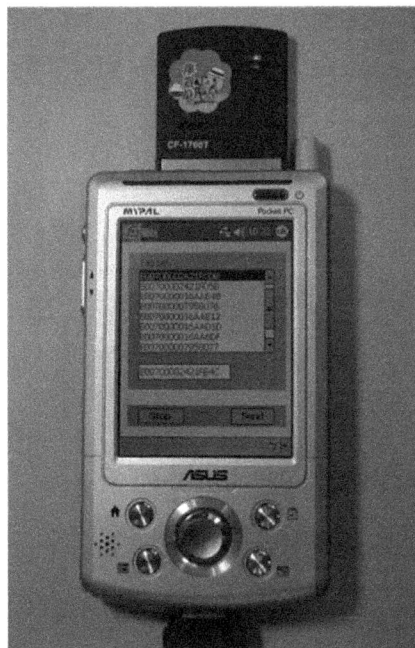

Fig. 4. The CF-card RFID reader mounted on he reader PDA.

standard. They can be detected by the reader with a reading/detection distance of 10 centimeters. An RFID reader with a CF (Compact Flash) transmission interface was used in this research. The reader model is CF-1700T by Sunion (http://www.sunion.com.tw/) with a working frequency at 13.56 MHz. The reader needs to be attached to a PDA with a CF slot. The power of the reader is supplied by the PDA with a consumption rate at 15 mA (max) in operation and 4 mA in idle. The PDA used for the RFID reader is of ASUS A716 Series with an Intel Xscale PXA255 400 MHz processor. It runs on a Microsoft Windows Powered Pocket PC 2003. The PDA has a CF Type II slot to plug-in the RFID reader and can access wireless LAN through WiFi (Fig. 4).

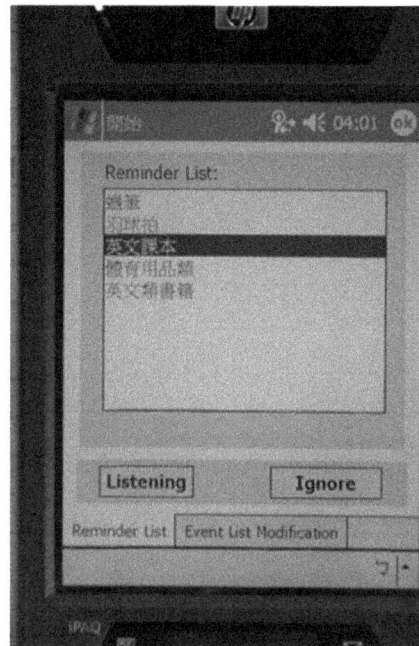

Fig. 5. The reminder list sent by the server is displayed on the client PDA.

In the application scenario of our system, the detection range should be from 0 to 100 cm or so. However, the RFID tags we used have a maximum detection distance of 10 cm. Thus in our tests, we had to put all the object tags close to the reader one by one. This can be corrected in real applications by using a doorway RFID portal. The object data along with their corresponding RFID tag IDs are saved in a database built by MS SQL Server 2005. With this setting, the objects taken by the user can be automatically detected. The PDA then transmits the detected RFID tag IDs to the reminder server via a wireless LAN. After a reminder list is generated, the server sends the list to the client PDA carried by the user (Fig. 5). The client PDA used in this research is HP iPAQ hx7400 with an Intel PXA270 624MHz processor. The operating system on the client PDA is Microsoft Windows Mobile 2003 Second Edition.

On the server side, the system was developed by Visual Studio 2005 C#. There are three user interfaces for system administration. The first one is for rule testing. The system administrator can select objects from the database for testing the correctness of the rules. Tag IDs from the reader PDA are not needed. The second one is for database management, including user data, object data and rule sets. The third one is for server management including initialization of the server and the communication between the server and the two PDAs (the reader PDA and the client PDA).

Experiments have been carried out with our prototype system. The user was assumed to be a school kid. A class schedule and some events were input into the system. Objects related to the classes as well as other personal belongings for daily activities were attached with an RFID tag. We then simulated the "going out" actions for one week. With the collected data, the objects are classified into one of the three frequency levels. The system then started to generate the reminder list to the user in the subsequent tests. In Fig. 5, when the user pushes the "Listening" button on the client PDA, the PDA receives and displays the reminder list generated by the server. Three objects and two object classes are displayed in one of our tests. The first three items are crayons, badminton racquet, and English textbook; the last two are sports goods and English books. The user can give a feedback on any of the reminding items by pushing the

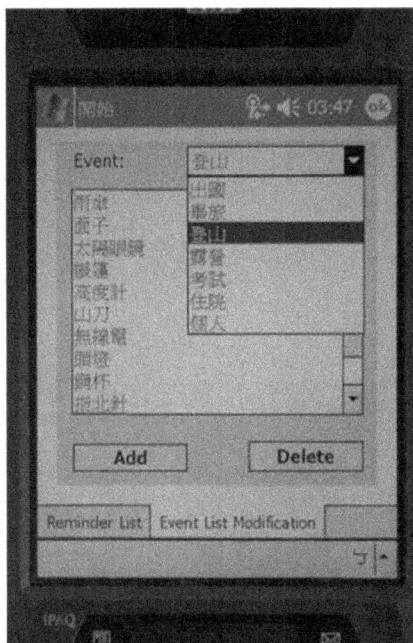

Fig. 6. Event objects editing.

"Ignore" button. The item will not appear in the same situation next time. In Fig. 6, the user can add or delete objects for an event on the client PDA. Objects for mountain climbing are shown in this example.

Generally speaking, our system could provide a personalized service in reminding the user what he/she had forgotten. The system is adaptive to the user's need. When the class schedule is changed for a new semester, it can be updated in the calendar. Also, the user can use the feedback mechanism to improve the system performance in reminding. However, in our tests, if the user brought things in an irregular manner, the system would not be able to provide useful reminder. In our experiments, the user was assumed to be a school kid. For further tests, people of different backgrounds and working professions should be considered. Next, in the current system, it can provide reminders to the user at weekends, but national holidays are not considered. Holidays can be treated as special events in the calendar. Moreover, the user may go out more than one time during a day. Time stamps can be added to the records of each "going out" and the system can analyze the user behavior not just based on the day of the week, but also the time of the day. Finally, the system can combine with weather forecast and event news from the Web. If the weather forecast says that the possibility of rain is very high this afternoon, the system can remind the user to take an umbrella with him/her. For example, if there are special discounts for members in a department store, the system can remind the user to bring the membership card.

## 5. Conclusion

A smart reminding mechanism is designed and presented in this paper. It utilizes RFID and mobile technologies. RFID provides an easy way to record the objects that the user takes out. Through the analysis on these history data, the system knows what the user should bring along on that specific day of the week. The analysis is based on the taken-out frequency of an object or a class of objects and a set of object/class rules. The event calendar is another source of information for determining what the user should bring along with. The objects for each event can be edited by the user or automatically generated by the system from the detected object list. After sensing the objects the user brings along with, the server constructs a reminder list with objects and classes and sends it to the client PDA or mobile phone carried by the user. The user then knows what items he/she might have forgotten. A prototype system has been developed in this research. With further testing and fine tuning, the system should be applicable to daily living.

## References

[1] V.B. Misic and J.V. Misic, Improving sensing accuracy in cognitive PANs through modulation of sensing probability, *Mobile Information Systems* **5** (2009), 177–193.

[2] W. Wu, X. Li, S. Xiang, H.-B. Lim and K.-L. Tan, Sensor relocation for emergent data acquisition in sparse mobile sensor networks, *Mobile Information Systems* **6** (2010), 155–176.

[3] E.M. Shakshuki, X. Xing and T.R. Sheltami, Fault reconnaissance agent for sensor networks, *Mobile Information Systems* **6** (2010), 229–247.

[4] H.-H. Hsu and C.-C. Chen, RFID-based human behavior modeling and anomaly detection for elderly care, *Mobile Information Systems* **6** (2010), 341–354.

[5] L. Jing, Z. Cheng, M. Kansen, T. Huang and S. Sun, An educational schoolbag system for providing an object reminder service, *IPSJ Digital Courier* **3** (2007), 64–74.

[6] Enjoy RFID technology. Obtained through the Internet: http://enjoyrfid.blogspot.com/, [accessed Feb. 7, 2011].

[7] J. Jayaputera and D. Taniar, Data retrieval for location-dependent queries in a multi-cell wireless environment, *Mobile Information Systems* **1** (2005), 91–108.

[8] A. Waluyo, B. Srinivasan and D. Taniar, Research on location-dependent queries in mobile databases, *International Journal of Computer Systems: Science and Engineering* **20** (2005), 79–95.

[9] A.B. Waluyo, D. Taniar, W. Rahayu and B. Srinivasan, Mobile service oriented architectures for NN-queries, *Journal of Network and Computer Applications* **32** (2009), 434–447.

[10] A.B. Waluyo, D. Taniar, W. Rahayu and B. Srinivasan, Mobile broadcast services with MIMO antennae in 4G wireless networks, *World Wide Web Journal* (2011). (DOI: 10.1007/s11280-011-0113-9, online for early access).

[11] A.B. Waluyo, W. Rahayu, D. Taniar and B. Srinivasan, A novel structure and access mechanism for mobile broadcast data in digital ecosystems, *IEEE Transactions on Industrial Electronics*, (2010). (DOI: 10.1109/TIE.2009.2035457, online for early access).

[12] Microsoft Home Virtual Presskit. Obtained through the Internet: http://www.microsoft.com/presspass/events/mshome/default.mspx, [accessed Feb. 7, 2011].

[13] Toyota Home. Obtained through the Internet: http://www.toyotahome.co.jp/, [accessed Feb. 7, 2011].

[14] *RFID Journal*. Obtained through the Internet: *http://www.rfidjournal.com/*, [accessed Feb. 7, 2011].

[15] J. Bohn, Prototypical implementation of location-aware services based on a middleware architecture for super-distributed RFID tag infrastructures, *Proceedings 19th International Conference on Architecture of Computing Systems* (ARCS06), 2006, pp. 155–166.

[16] Y. Nakauchi, T. Suzuki, A. Tokumasu and S. Murakami, Cooking procedure recognition and support system by intelligent environments, *Proceedings IEEE Workshop on Robotic Intelligence in Informationally Structured Space* (RIISS '09), 2009, pp. 99–106.

[17] M. Darianian and M.P. Michael, Smart home mobile RFID-based Internet-of-things systems and services, *Proceedings 2008 International Conference on Advanced Computer Theory and Engineering*, 2008, pp. 116–120.

[18] E. Becker, V. Metsis, R. Arora, J. Vinjumur, Y. Xu and F. Makedon, SmartDrawer: RFID-based smart medicine drawer for assistive environments, *Proceedings 2nd International Conference on PErvasive Technologies Related to Assistive Environments* (PETRA'09), 2009.

[19] M. Stikic, T. Huỳnh, K.V. Laerhoven and B. Schiele, ADL recognition based on the combination of RFID and accelerometer sensing, *Proceedings Second International Conference on Pervasive Computing Technologies for Healthcare* (PervasiveHealth 2008), 2008, pp. 258–263.

[20] S. Im, I.J. Kim, S.C. Ahn and H.G. Kim, Automatic ADL classification using 3-axial accelerometers and RFID sensor, *Proceedings IEEE International Conference on Multisensor Fusion and Integration for Intelligent Systems* (MFI 2008), 2008, pp. 697–702.

[21] D. Cook and S. Das, *Smart Environments: Technology, Protocols and Applications*, Wiley-Interscience, 2005.

[22] R. Kadouche, B. Abdulrazak, M. Mokhtari, S. Giroux and H. Pigot, A semantic approach for accessible services delivery in a smart environment," *International Journal of Web and Grid Services* **5** (2009), 192–218.

[23]  B. Abdulrazak, B. Chikhaoui, C. Gouin-Vallerand and B. Fraikin, A standard ontology for smart spaces, *International Journal of Web and Grid Services* **6** (2010), 244–268.
[24]  V.R. Jakkula, A.S. Crandall and D.J. Cook, Knowledge discovery in entity based smart environment resident data using temporal relation based data mining, *Proceedings Seventh IEEE International Conference on Data Mining Workshops* (ICDMW 2007), 2007, pp. 625–630.
[25]  P. Rashidi and D.J. Cook, Keeping the intelligent environment resident in the loop, *Proceedings 4th International Conference on Intelligent Environments*, (IE 2008), 2008, pp. 1–9.
[26]  D.J. Cook, M.S. Edgecombe, A. Crandall, C. Sanders and B. Thomas, Collecting and disseminating smart home sensor data in the CASAS project, *Proceedings CHI Workshop on Developing Shared Home Behavior Datasets to Advance HCI and Ubiquitous Computing Research*, 2009.
[27]  O. Brdiczka, J.L. Crowley and P. Reignier, Learning situation models in a smart home, *IEEE Transactions on Systems, Man and Cybernetics – Part B* **39** (2009), 56–63.
[28]  H. Si, S.J. Kim, N. Kawanishi and H. Morikawa, A context-aware reminding system for daily activities of dementia patients, *Proceedings 27th International Conference on Distributed Computing Systems Workshops* (ICDCSW'07), 2007, p. 50.

**Hui-Huang Hsu** is an Associate Professor in the Department of Computer Science and Information Engineering at Tamkang University, Taipei, Taiwan. He received his PhD and MS Degrees from the Department of Electrical and Computer Engineering at the University of Florida, USA, in 1994 and 1991, respectively. He has published over 90 referred papers and book chapters, as well as participated in many international academic activities. His current research interests are in the areas of machine learning, data mining, ambient intelligence, bio-medical informatics, and multimedia processing. Dr. Hsu is a senior member of the IEEE.

**Cheng-Ning Lee** received his MS Degree from the Department of Computer Science and Information Engineering at Tamkang University, Taipei, Taiwan in 2010. His research interests include mobile computing and RFID applications.

**Jason C. Hung** is an Associate Professor of Department of Information Management at Overseas Chinese University, Taiwan, ROC. His research interests include Multimedia Computing and Networking, Distance Learning, E-Commerce, and Agent Technology. From 1999 to date, he was a part time faculty of the Computer Science and Information Engineering Department at Tamkang University. Dr. Hung received his BS and MS degrees in Computer Science and Information Engineering from Tamkang University, in 1996 and 1998, respectively. He also received his Ph.D. in Computer Science and Information Engineering from Tamkang University in 2001. Dr. Hung has published over 70 papers and book chapters, as well as participated in many international academic activities, including the organization of many international conferences. He is the founder and Workshop chair of International Workshop on Mobile Systems, E-commerce, and Agent Technology. He is also the Associate Editor of the International Journal of Distance Education Technologies, published by Idea Group Publishing, USA. Web: http://www.ocu.edu.tw ⌐jhung.

**Timothy K. Shih** is a Professor of Department of Computer Science and Information Engineering, National Central University, Taiwan. He is a Fellow of the Institution of Engineering and Technology (IET). In addition, he is a senior member of ACM and a senior member of IEEE. He has edited many books and published over 430 papers and book chapters. He was the founder and co-editor-in-chief of the International Journal of Distance Education Technologies, published by Idea Group Publishing, USA. He is an Associate Editor of the ACM Transactions on Internet Technology and an associate editor of the IEEE Transactions on Learning Technologies. He was also an Associate Editor of the IEEE Transactions on Multimedia. He has received many research awards, including research awards from National Science Council of Taiwan, IIAS research award from Germany, HSSS award from Greece, Brandon Hall award from USA, and several best paper awards from international conferences.

# 10

# A distributed approach to continuous monitoring of constrained k-nearest neighbor queries in road networks

Hyung-Ju Cho*, Seung-Kwon Choe and Tae-Sun Chung
*Department of Computer Engineering, Ajou University, Gyeonggi-Do, South Korea*

**Abstract.** Given two positive parameters $k$ and $r$, a constrained k-nearest neighbor (CkNN) query returns the $k$ closest objects within a network distance r of the query location in road networks. In terms of the scalability of monitoring these CkNN queries, existing solutions based on central processing at a server suffer from a sudden and sharp rise in server load as well as messaging cost as the number of queries increases. In this paper, we propose a distributed and scalable scheme called DAEMON for the continuous monitoring of CkNN queries in road networks. Our query processing is distributed among clients (query objects) and server. Specifically, the server evaluates CkNN queries issued at intersections of road segments, retrieves the objects on the road segments between neighboring intersections, and sends responses to the query objects. Finally, each client makes its own query result using this server response. As a result, our distributed scheme achieves close-to-optimal communication costs and scales well to large numbers of monitoring queries. Exhaustive experimental results demonstrate that our scheme substantially outperforms its competitor in terms of query processing time and messaging cost.

Keywords: Distributed algorithm, continuous monitoring, constrained k-NN query, road network

## 1. Introduction

The increased popularity of mobile communication devices with embedded positioning capabilities (e.g., GPS) has triggered the development of many location-based applications. The ability to support location-based queries from mobile clients on a road network is essential for this class of mobile applications.

In this work, we investigate the continuous monitoring of $k$ nearest neighbor queries over static objects within a query distance $r$ of the current location where the $k$ and $r$ values are provided by the clients. A location based query that reports the $k$ nearest neighbors within a query distance $r$ of a moving query point $q$ is called a constrained $k-$nearest neighbor (CkNN) query [8,9,12]. These queries are often useful to find specific objects (e.g., gas stations reachable in 10 minutes driving) within some specific region. Consider the example of a family traveling by car. They may want to continuously monitor the three closest restaurants within 5 km of their current location so that they can choose a restaurant that serves their favorite food.

This paper assumes that (1) query points move freely on the road network and any mobility pattern (e.g., speed, direction, route, etc.) on them is not given, (2) objects (e.g., gas stations, restaurants) of

*Corresponding author: Department of Computer Engineering, Ajou University Woncheon-dong, Suwon Si Yeongtong-gu, Gyeonggi-Do, 443-749, South Korea. E-mail: hjcho@ajou.ac.kr.

interest to the clients do not change their location on the road network, and (3) a central server and each client talk to each other using a wireless channel such as cellular services or Wi-Fi. Last, like most on-line monitoring systems, main-memory evaluation is exploited.

Continuous monitoring of $k$-nearest neighbor queries and range queries on the road network has been well studied [5,14,18,21,23,26,27,29]. In real-life scenarios where (1) a large and rapidly growing number of monitoring queries are handled and (2) complex queries are processed, existing solutions which rely on central processing at the server suffer from high server loads and messaging costs [3,10, 22]. We address this issue by proposing a *d*istributed and scalable scheme called DAEMON which is tailored for the continuous *mon*itoring of C$k$NN queries in the road network.

Our distributed processing of C$k$NN queries consists of server side processing and client side processing. The server is dedicated to efficiently processing C$k$NN queries issued at intersections and quickly retrieving the objects in the road segments between neighboring intersections. The clients are devoted to making their own query result using the information provided by the server. Our proposed scheme is distinct from its rivals which largely depend on the capabilities of a centralized server. First, in our scheme, clients do not register their continuous queries to the server. Second, the clients do not send their location information to the server, which helps ease privacy issues regarding the locations of users [1, 19]. Last, the server burden is dramatically decreased by shipping part of the query processing to the clients [3,10,22].

At the server side, a shared execution technique is employed to decrease the processing time of queries issued at intersections of road segments. Among all nodes, a small portion of them, which we call condensing nodes, are carefully selected using query results. Each condensing node keeps information on the objects within an observation distance $R$. Finally, using the server response, clients generate the query results for the road segments which they are currently at.

Our principal contributions are summarized as follows:

- We present a distributed and scalable scheme called DAEMON for continuous monitoring of C$k$NN queries in road networks.
- Our scheme minimizes server load and messaging cost by shipping part of the query processing to the clients. The server neither exploits nor stores movement information (e.g., route or location updates) of the clients. This enables the server to run this service on-demand and mitigates location privacy issues.
- We employ the shared execution technique called condensing nodes to reduce the processing time at the server. This leads to a significant reduction in the evaluation time of queries issued at intersections of road segments.
- We conduct an exhaustive performance evaluation to show that DAEMON is markedly superior to its competitor under various conditions while achieving close-to-optimal communication costs.

The rest of the paper is structured as follows: Section 2 briefly reviews related work on $k$ NN algorithms and query monitoring. Section 3 presents preliminaries and formulates the problem for this work. Sections 4 to 6 elaborate on our distributed approach to continuous monitoring of C$k$NN queries on the road network. Thorough experimental results are given in Section 7. Finally, Section 8 concludes the paper.

## 2. Related work

For location-based services, processing $k$-nearest neighbor ($k$NN) queries on the road network has been well studied (e.g. [1,5,6,8,9,12,14,18,20,27,28]). For a good survey on location based query processing, refer to [14]. Papadias et al. proposed the incremental network expansion (INE) algorithm for $k$NN queries. The basic idea of this algorithm is to incrementally search the road segments from the query point until the $k$-nearest objects of the query point are found [20]. The performance of the INE algorithm is closely related to the density of objects in the network. Hence, the INE suffers dramatic degradation in performance when objects are sparsely distributed. To overcome this limitation, many $k$NN query processing algorithms which utilize the pre-computed network distance to optimize the query processing (e.g. [1,6,27]) have been proposed. Bao et al. introduced a new type of query called a $k$-range nearest neighbor ($k$RNN) query to find the $k$ closest objects of every point on the road segments inside a given query region based on the network distance [1]. They presented an algorithm which exploits a shared execution paradigm to eliminate the redundant searching overhead among adjacent nodes. Ghadiri et al. presented two fuzzy clustering methods for group nearest neighbor queries based on fuzzy logic distance model [11].

There is a large body of research work on continuous monitoring of spatial queries in Euclidean spaces. Cheema et al. present a safe zone based approach to continuous monitoring of distance-based range queries that return the objects within a distance $r$ of the query location [4]. Mokbel et al. proposed an algorithm called SINA for evaluating a set of concurrent spatial queries in order to reduce the overall computational cost using shared execution and incremental evaluation [15]. Distributed approaches have also been investigated to monitor continuous range queries [3,10] and continuous $k$NN queries [22]. The main idea of these distributed schemes is to shift some of the load from the server to mobile clients. Morvan et al. proposed a mobile relational algebra to decentralize the control of dynamic query optimization processes [17]. Hasan et al. presented algorithms which are applicable to any arbitrarily-shaped constrained region for continuous monitoring of C$k$NN queries [12] and Gao et al. presented algorithms for processing C$k$NN searches on the trajectories of moving objects [9]. However, the techniques based on Euclidean distance are not applicable to our problem concerning network-constrained mobile clients and network distance-based queries.

There is also a large body of research on continuous monitoring of spatial queries for road networks. Chen et al. dealt with path $k$-NN queries that return $k$NNs with respect to the shortest path connecting the destination and the user's current location [5]. Stojanovic et al. proposed a technique for continuous monitoring of range queries over moving objects [21]. The range of the query may be defined by a user selected area, a map window, a polygon, a circle or a part of a road segment. Recently, Xuan et al. proposed several Voronoi-based algorithms for continuous range queries [23–25] as well as continuous $k$ NN queries [27]. Their algorithms may suffer from high computational cost at the server and high communication cost as the number of monitoring clients increases rapidly.

Mouratidis et al. presented two algorithms, IMA and GMA, for the continuous nearest neighbor monitoring problem in road networks. In this problem, both queries and objects of interest move freely [18]. GMA integrates IMA with the shared execution paradigm. Specifically, GMA groups together the queries falling on the path between two consecutive intersections in the network and monitors the $k$-NNs of the intersections. It then utilizes query results at the intersections in order to facilitate evaluation of the queries on the path. It has been shown that GMA is typically better than IMA. Therefore, we focus mainly on GMA in the paper.

Even if GMA and IMA can be extended to continuously monitor C$k$NN queries over static objects, they are not very efficient because they are not specifically designed for monitoring these C$k$NN queries. GMA demonstrates the following disadvantages in terms of scalability. (1) The query processing of GMA is centralized at the server, which is in sharp contrast to our distributed scheme. Hence, in GMA, the server burden grows rapidly with the number of monitoring queries. (2) GMA requires that clients install their continuous queries on the server and report their changed location periodically in order for the server to capture the status (e.g., location and query conditions) of each client. None of both are demanded by our scheme. These requirements may raise location privacy problem. (3) Frequent location updates greatly increase the server load. Whenever clients move and their movements are reported to the server, the queries are re-evaluated at the server. If query results are updated, the updated results then need to be delivered to the clients through a wireless network. In addition, the server may not be able to cope with the high location report rate which is necessary to ensure accurate answers.

## 3. Preliminaries

In this section, we describe the road network, system assumptions, and terms used in the paper and present the formal definition of our problem.

### 3.1. Road network and system assumptions

We model the underlying road network as a weighted undirected graph $G = (N, E)$ where $N$ is a finite set of nodes, $E$ is a set of edges (i.e., road segments), and $E \in N \times N$. Each edge is given the length of its corresponding road segment as a weight.

In this work, we consider our system to be a mobile environment where the mobile clients can communicate with a central server through some wireless communication infrastructure (e.g., cellular services or Wi-Fi). The clients are equipped with positioning technology like GPS and can locate their positions. They also have sufficient capabilities to carry out computational tasks. All these assumptions are either widely agreed upon or are seen as common practice in most existing mobile systems in the context of the monitoring and tracking of moving objects [3,7,10,13,16,22].

### 3.2. Definitions of terms and problem

To clarify the meanings of the primary terms used in this paper, we present the terms and their formal definitions. Nodes can be categorized according to node degree. (1) If a node degree is larger than 2, the node is referred to as an intersection node. (2) If a node degree is 2, the node is referred to as an intermediate node. (3) If a node degree is 1, the node is referred to as a terminal node. A certain amount of intersection nodes are chosen and referred to as condensing nodes. Each condensing node is used to store the information (i.e., object's id, location, and distance to the condensing node) about the objects that lie within its observation distance $R$.

A sequence denotes a path between two nodes $n_s$ and $n_e$, such that $n_s$ (or $n_e$) is either an intersection node or a terminal node and the other nodes in the path are intermediate nodes. Two end nodes $n_s$ and $n_e$ of a sequence are called the boundary nodes. Particularly, the sequence where a query point $q$ remains is called the (current) active sequence of $q$. Table 1 presents the primary symbols used in this paper and their definition.

Using the network in Fig. 1, we formulate our continuous monitoring problem. In this figure, objects $a$, $b$, $c$ represent entities of interest and the query point $q$ corresponds to a client that requests continuous monitoring of the $k$ closest objects within a query distance $r$ of its current location. The numbers above the road segments indicate the network distances of two adjacent objects or nodes. That is, $d(n_2, a) = d(n_6, c) = 1.5$ and $d(n_2, b) = d(b, n_5) = 1$. Suppose that $k = 2$ and $r = 1$ are given and $q$ moves along sequence $SQ_{(n_2, n_5, n_6)}$. Then, while $q$ runs between $n_2$ and $n_5$, the C$k$NN query result

Table 1
Primary symbols and their definition

| Symbol | Definition |
|---|---|
| $SQ_{(n_i,n_{i+1},...,n_j)}$ | A sequence where $n_s$ (or $n_e$) is the start (or end) boundary node and the others $n_{s+1}, \ldots, n_{e-1}$ are intermediate nodes. Start and end boundary nodes are decided by the movement direction of query point |
| $d(a,b)$ | The shortest path length between object $a$ and object $b$ |
| $q$ | Query point on the road network (e.g., the current location of a vehicle) |
| $Aq$ | The set of objects satisfying query conditions at location $q$ |
| $k$ | Number of nearest neighbors requested by C$k$NN query |
| $r$ | Network distance requested by C$k$NN query |
| $R$ | Observation distance of condensing node |

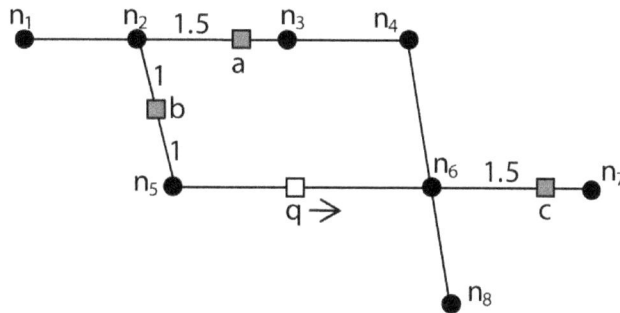

Fig. 1. Simple road network for problem definition.

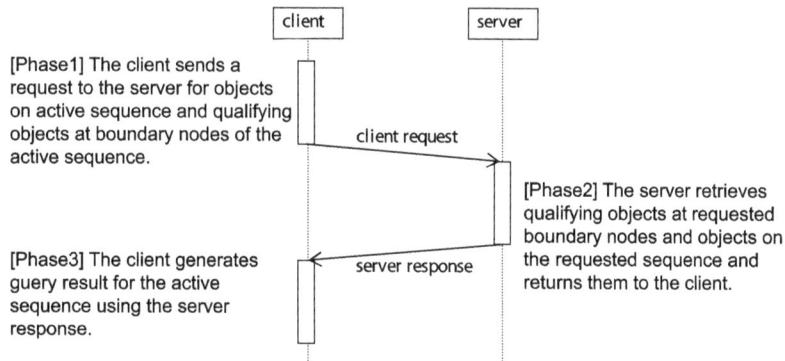

Fig. 2. Interaction diagram between client and server.

is object $b$ and while $q$ runs between $n_5$ and $n_6$, the query result is empty. Since we do not assume any knowledge about the movement patterns of queries, the next active sequence is not revealed before the query point actually gets to it. On reaching $n_6$ which is the end of the current active sequence $SQ_{(n_2,n_5,n_6)}$, $q$ may choose any of the next active sequence candidates $SQ_{(n_6,n_4,n_2)}$, $SQ_{(n_6,n_7)}$, $SQ_{(n_6,n_8)}$.

## 4. Continuous monitoring of constrained $k$ NN queries

Section 4.1 gives an overview of our scheme which consists of three phases as illustrated in Fig. 2. Sections 4.2, 5, and 6 elaborate on the first, second, and third phase, respectively.

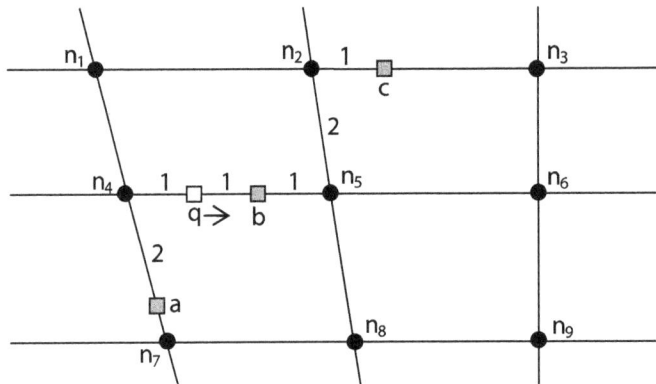

Fig. 3. Example road network and CkNN query $q$ with $k = 2$ and $r = 3$.

### 4.1. Overview

Figure 2 depicts the interaction diagram between the client and server. It is important to note that while a client travels along the active sequence, an interaction which consists of the client request and server response is sufficient for the client to generate the query result for the active sequence. Naturally, this interaction is repeated whenever the client reaches a new active sequence. To simplify the presentation, we focus on the interaction between a client and the server while the client runs on the active sequence. The three phases of Fig. 2 are fully described in Sections 4.2, 5, and 6, respectively.

In the first phase, the client asks the server for the objects on the active sequence and the qualifying objects which meet query conditions at each boundary node of the active sequence. Note that the client does not provide its location information to the server.

In the second phase, the server processes the client request and returns the response through the wireless network. The request consists of the following two tasks: (1) evaluating CkNN queries issued at the boundary nodes of the active sequence and (2) retrieving the objects on the active sequence. After the processing at the server is complete, the processed result is returned to the client over the wireless channel.

In the final phase, using the server response, the client generates the query result for the active sequence. The query result consists of valid intervals and their corresponding qualifying objects where a valid interval indicates a part of the travel path where a set of qualifying objects remains unchanged.

The characteristics of our proposed architecture can be summarized as follows: First, the server is tailored to efficiently evaluate CkNN queries issued at boundary nodes and to rapidly retrieve the objects on the active sequences. The clients are devoted to quickly generating their query results using the server responses. Such distributed processing greatly reduces the server burden and communication costs. Second, each client sends only one request to the server while on a particular active sequence. This achieves satisfactory communication costs between the clients and the server in most cases. Last, the clients do not disclose their location information to the server, which alleviates privacy issues regarding location information.

### 4.2. Sending client request to server

In the rest of the paper, we utilize the example road network in Fig. 3 to explain the three phases of Fig. 2. In Fig. 3, query $q$ with $k = 2$ and $r = 3$ moves along the active sequence $SQ_{(n_4, n_5)}$.

The numbers above the road segments represent the distances between two adjacent objects. That is, $d(n_4, a) = d(n_4, b) = d(n_2, n_5) = 2$ and $d(n_5, b) = d(n_2, c) = 1$.

Each client request is in a form $< qid, k, r, boundary\_node(s), sequence >$ where $qid$ indicates query id (e.g., vehicle id), $k$ and $r$ indicate query conditions detailed in Table 1, $boundary\_node(s)$ represent boundary nodes of the active sequence, and the last attribute $sequence$ represents the active sequence.

Recall that the exemplary C$k$NN query $q$ of Fig. 3 requests continuous monitoring of two NNs within three (kilometers) of its location. The first request $< q, 2, 3, \{n_4, n_5\}, SQ_{(n_4, n_5)} >$ is generated and transferred to the server. Soon, the client receives the response from the server and makes its own query result for $SQ_{(n_4, n_5)}$. In the case where the client turns left at boundary node $n_5$, the second request $< q, 2, 3, \{n_2\}, SQ_{(n_5, n_2)} >$ is transferred to the server. Observe that $n_5$ is not included in the fourth attribute (i.e., $\{n_2\}$) of the second request. The reason for this is that the response for the previous active sequence $SQ_{(n_4, n_5)}$ is cached at the client and recycled. Specifically, the information (i.e., object's id, location, and distance to $n_5$) on the qualifying objects at $n_5$ is re-used by the client.

## 5. Processing client request at the server

Section 5.1 describes our C$k$NN search algorithm which exploits condensing nodes and sequences. Section 5.2 discusses the updates of objects such as insertion and deletion of objects.

### 5.1. CkNN search algorithm

Upon receiving the client request, the server starts to process the request. Recall that the client request consists of two tasks. One is to evaluate queries issued at the boundary nodes of the active sequence and the other is to retrieve the objects on the active sequence. To process the client request effectively, we introduces condensing nodes and sequences both of which significantly reduce the computation time of the server. Condensing nodes and sequences are stored in the condensing node table CT and the sequence table ST, respectively. Both of CT and ST are hash tables on node id and sequence id, respectively.

Condensing nodes are generated and updated using query results at runtime. Each condensing node stores information about the objects within its observation distance $R$. This results in reducing the network expansion from the condensing node. As a result, the condensing nodes enable the server to quickly evaluate queries issued at boundary nodes. Each entry in condensing node table CT is in a form $<nid, loc, R, O, adj\_seq>$ where $nid$ is a node id for a condensing node $n_c$, $loc$ indicates the position of $n_c$, $R$ denotes the observation distance of $n_c$, $adj\_seq$ denotes the set of adjacent sequences of $n_c$ and $O$ denotes the set of pairs $< obj, d(n_c, obj) >$ where $obj$ is an object such that $d(n_c, obj) \leqslant n_c.R$.

Each sequence has the information on the objects on the sequence. We note that the total number of sequences is typically smaller than that of edges in the network since a sequence may include multiple edges. Each entry in sequence table ST is in a form $< seqid, (n_s, n_{s+1}, \ldots n_e), O_{(n_s, n_{s+1}, \ldots n_e)}, len >$ where $seqid$ is a sequence id for $SQ_{(n_s, n_{s+1}, \ldots, n_e)}$, $n_s$ and $n_e$ represent the boundary nodes, $(n_s, n_{s+1}, \ldots n_e)$ is a list of ids of nodes contained in the sequence, $O_{(n_s, n_{s+1}, \ldots n_e)}$ represents the set of pairs $< obj, d(n_s, obj) >$ where $obj$ is an object on $SQ_{(n_s, n_{s+1}, \ldots, n_e)}$, and $len$ is the sequence length which is the total weight of edges in the sequence. The server can quickly retrieve the objects on the requested sequence by using ST. In the rest of this section, we describe our C$k$NN search algorithm to rapidly evaluate queries issued at boundary nodes.

The C$k$NN search algorithm does two jobs. One is to evaluate a C$k$NN query and the other is to update with query result the information of the boundary node $n_q$ which corresponds to query location

---

**Algorithm 1:** C$k$NN _search $(q,q.k,q.r)$

**Input:** $q$: a query point, $q.k$: the number of NNs, $q.r$: query distance
**Output:** a set of objects satisfying query predicate and their distance to $q$
1:    $kth\_dist \leftarrow q.r$    /*$kth\_dist$ is initialized to the $q.r$ value */
2:    $q.A \leftarrow \phi$    /* answer set $q.A$ keeps the objects satisfying query predicate */
3:    $PQ.push(q,0)$    /*$PQ$ is a priority queue. note that $d(q,q)=0$ */
4:    **while** $PQ$ is not empty **do**
5:        $(n_{top}, d(q,n_{top})) \leftarrow PQ.$pop()
6:        Mark $n_{top}$ as visited in order to avoid multiple visits.
7:        **if** $d(q,n_{top}) \leqslant kth\_dist$ **then**
8:            search_boundary_node $(n_{top}, d(q,n_{top}), kth\_dist)$
9:        **else** /* it means that $d(q,n_{top}) > kth\_dist$*/
10:          exit while loop
11:   /* recall that $n_q$ is the boundary node which corresponds to $q$ */
12:   **if** $n_q$ is an intersection node **and** $kth\_dist > n_q.R$ **then**
13:        **if** $kth\_dist \leqslant R_{\max}$ **then**      $n_q.O \leftarrow q.A, \quad n_q.R \leftarrow kth\_dist$
14:        **else** /* it means that $kth\_dist > R_{\max}$ */
15:          $n_q.R \leftarrow R_{\max}$   /* note that $n_q.R$ is set to $R_{\max}$ */
16:          **for** each object $o \in q.A$ **do**
17:            **if** $d(n_q,o) \leqslant R_{\max}$ **then**    $n_q.O \leftarrow n_q.O \cup \{(o,d(n_q,o))\}$
18:   **return** $q.A$

---

if necessary. Note that query evaluation is described in lines 1 to 10 of the algorithm and boundary node update is described in lines 11 to 17.

The distance $kth\_dist$ between $q$ and its $k$th NN $o_{kth}$ is initialized to the query distance $q.r$ value. First, the tuple $(q,0)$ is pushed to the priority queue $PQ$ for a network expansion. Note that a tuple of $PQ$ consists of a node to be searched and its distance to $q$. If $PQ = \phi$ or $d(q,n_{top}) > kth\_dist$, the algorithm exits the while loop. Otherwise, $n_{top}$ and its adjacent sequences are searched iteratively.

If the boundary node $n_q$ is an intersection node and $n_q.R < kth\_dist$, $n_q.R$ and $n_q.O$ are updated using $kth\_dist$ and query answer $q.A$, respectively. Note that terminal nodes and intermediate nodes cannot be condensing nodes. Observation distance of each boundary node is initialized to 0 and cannot exceed the maximum observation distance $R_{\max}$ value which is given as a system parameter. If $n_q.R < kth\_dist \leqslant R_{\max}$, $n_q.R$ and $n_q.O$ are updated to $kth\_dist$ and $q.A$, respectively. If $kth\_dist > R_{\max}$, $n_q.R$ is set to the $R_{max}$ value and $n_q.O$ is updated to the set of objects $o$ such that for each object $o \in q.A$, $d(q,o) \leqslant R_{\max}$. Condensing nodes are kept in memory and sorted in a least recently used (LRU) list. The last entry is removed if there is no space available to accommodate a new condensing node.

The Search_boundary_node algorithm explores the objects within the observation distance of $n_{top}$ and its adjacent sequences with an offset $d(q,n_{top})$. If $r$ is a condensing node, the algorithm first investigates each object $o$ in $n_{top}.O$. If $d(q,n_{top}) + d(n_{top},o) \leqslant kth\_dist$, a tuple $(o,d(q,o))$ is added to answer set $q.A$ and $kth\_dist$ is updated to the value of $d(q,o_{kth})$ if $o_{kth} \in q.A$. Next, each adjacent sequence of $r$ is explored. If $n_{top}$ is a condensing node and its observation distance is not smaller than the length of an adjacent sequence of $n_{top}$, the adjacent sequence is skipped since the objects on the sequence have already been investigated. Otherwise, the adjacent sequence should be explored. Finally, among adjacent boundary nodes of $r$, the nodes which have not been explored are identified and are pushed to $PQ$.

The Search_sequence algorithm examines the objects on an adjacent sequence of $n_{top}$. For each object $o$ on the adjacent sequence, we check if $d(q,o) \leqslant kth\_dist$. If so, a tuple $(o,d(q,o))$ is added to $q.A$ and $kth\_dist$ is updated to the value of $d(q,o_{kth})$. When the C$k$NN search algorithm ends, the qualifying objects in $q.A$ and their information (e.g., location and distance to $q$) are returned to the client.

---

**Algorithm 2:** Search_boundary_node $(n_{top}, d(q, n_{top}), kth\_dist)$

---

**Input:** $n_{top}$: a node popped from $PQ$, $d(q, n_{top})$: an offset, $kth\_dist$: distance from $q$ to its $k$th NN
**Output:** a set of objects satisfying query predicate at $n_{top}$ and its adjacent sequences
1:   /* let $n_{top}.O$ be the set of objects within observation distance of $n_{top}$ */
2:   **if** $n_{top}$ is a condensing node **then**  /* if $n_{top}.R > 0$, $n_{top}$ is a condensing node */
3:     **for** each object $o \in n_{top}.O$ **do** /* if $n_{top}$ is a condensing node, each object in $n_{top}.O$ is examined */
4:       **if** $d(q, n_{top}) + d(n_{top}, o) \leqslant kth\_dist$ **then**
5:         $q.A \leftarrow q.A \cup \{(o, d(q, n_{top}) + d(n_{top}, o)\}$
6:         $kth\_dist \leftarrow d(q, o_{kth})$ where $o_{kth} \in q.A$
7:     **for** each adjacent sequence $SQ_{(n_{top}, n_j)} \in n_{top}.adj\_seq$ **do**
8:       **if** $n_{top}$ is a condensing node **and** $SQ_{(n_{top}, n_j)}.len \leqslant n_{top}.R$ **then**
9:         skip the visit to $SQ_{(n_{top}, n_j)}$
10:      **else**
11:        Search_sequence $(SQ_{(n_{top}, n_j)}, d(q, n_{top}), kth\_dist)$
12:        **if** $n_j$ is not marked as visited **then**  $PQ$.push $(n_j, d(q, n_{top}) + d(n_{top}, n_j))$
13:   **return** $q.A$

---

**Algorithm 3:** Search_sequence $(SQ_{(n_{top}, n_j)}, d(q, n_{top}), kth\_dist)$

---

**Input:** $SQ_{(n_{top}, n_j)}$: an adjacent sequence of $n_{top}$, $d(q, n_{top})$ and $kth\_dist$ have the same meanings as before
**Output:** a set of objects satisfying query predicate on an adjacent sequence of $n_{top}$
1:   **for** each object $o \in O_{(n_{top}, n_j)}$ **do**
2:     **if** $d(q, n_{top}) + d(n_{top}, o) \leqslant kth\_dist$ **then**
3:       $q.A \leftarrow q.A \cup \{(o, d(q, o))\}$
4:       $kth\_dist \leftarrow d(q, o_{kth})$ where $o_{kth} \in q.A$
5:   **return** $q.A$

---

## 5.2. Discussion on updates of objects

Finally, we discuss the updates in the set of objects even if the updates occur rarely. There are three types of updates to a set of objects: an object addition (e.g., opening of a gas station), an object deletion (e.g., closing of a gas station), and an object movement. Object movement can be treated as the deletion and addition of the object. Such changes are easily reflected to the CT and ST. Consider that a new object $o_{new}$ is added. It is necessary that the sequence and condensing nodes which are affected by the addition of $o_{new}$ are identified. To do this, we simply find the sequence including $o_{new}$ and the condensing nodes within the maximum observation distance $R_{max}$ from $o_{new}$. Finally, new entries for the object $o_{new}$ are appended to the affected sequence and condensing nodes. The object deletion is handled in a similar way to the object addition.

## 6. Generating query result for active sequence at the client

Section 6.1 describes the distance graph between a moving query point and an object. Section 6.2 elaborates on decision of valid intervals and their query answer.

### 6.1. Distance graph between moving query point and qualifying object

Using the response from the server, the client generates valid intervals and their answer for the active sequence. We continue to use the example of Fig. 3. Now the client has query results (i.e., $A_{n_4}$ and $A_{n_5}$) at both boundary nodes of the active sequence and the information on the objects (i.e., $O_{(n4, n5)}$) on the active sequence. Specifically, $A_{n_4} = \{< obj, d(n_4, obj) > | q(obj, n_4) \leqslant 3\} = \{< a, 2 >,$

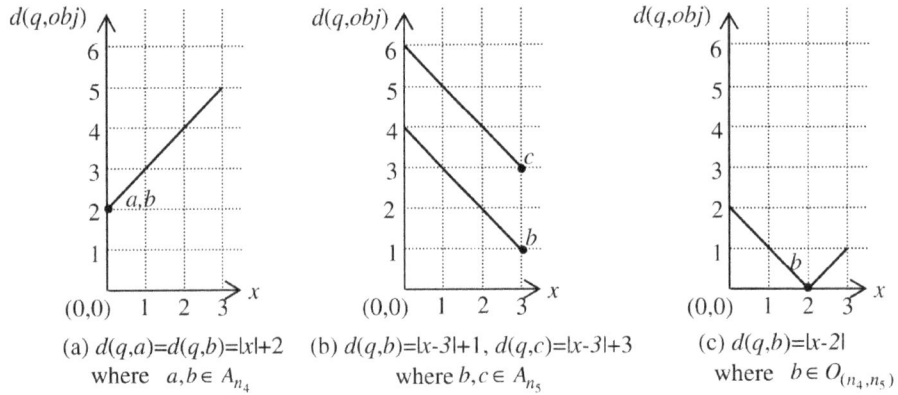

(a) $d(q,a)=d(q,b)=|x|+2$  (b) $d(q,b)=|x-3|+1$, $d(q,c)=|x-3|+3$  (c) $d(q,b)=|x-2|$
   where  $a,b \in A_{n_4}$        where $b,c \in A_{n_5}$           where  $b \in O_{(n_4,n_5)}$

Fig. 4. Distance graph between $q$ and each of $a, b, c$.

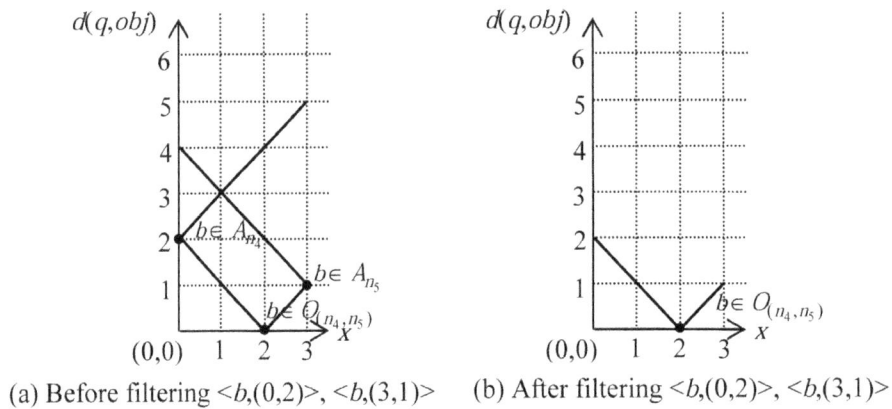

(a) Before filtering $<b,(0,2)>$, $<b,(3,1)>$    (b) After filtering $<b,(0,2)>$, $<b,(3,1)>$

Fig. 5. Filtering redundant tuples $< b, (0, 2) >$, $< b, (3, 1) >$ for object $b$.

$< b, 2 >$}, $A_{n_5} = \{< obj, d(n_5, obj) > |d(obj, n_5) \leqslant 3\} = \{< b, 1 >, < c, 3 >\}$, and $O_{(n4,n5)} = \{<$ $obj, d(n_4, obj) > |obj \in SQ_{(n_4,n_5)}\} = \{< b, 2 >\}$.

Figure 4 depicts the change of the network distance between query point $y_2 \geqslant |x_2 - x_1| + y_1$ and each of objects $a$, $b$, $c$ while $q$ moves on $SQ_{(n_4,n_5)}$. In these graphs, the $x$ and y values indicate $d(q, n_4)$ and $d(q, obj)$, respectively, where $obj \in A_{n_4} \cup A_{n_5} \cup O_{(n_4,n_5)}$. For instance, $x = 2$ means that $q$ reaches object $b$. In Figs 4 to 7, each bold dot (e.g., (0,2), (3,3), (3,1), (2,0) in Fig. 4) represents an apex point of a piecewise-linear graph. At the apex point, the distance between $q$ and an object is smallest. In other words, as $q$ is far from the apex point of an object, the distance of $q$ to the object increases. For simplicity, we consider the edges to be bidirectional, but our methods can be easily applied to networks with unidirectional edges (e.g., one way roads or roads where the weight is not same for different directions).

From Fig. 4, one can see that while $q$ moves on a sequence $SQ_{(n_s,n_{s+1},...n_e)}$, the distance between $q$ and each object in $A_{n_s}$, $A_{n_e}$, or $O_{(n_s,n_{s+1},...,n_e)}$ is represented by a piecewise-linear graph. Therefore, it is necessary to determine the coordinate of the apex point of the piecewise-linear graph. The coordinate $(x_{apex}, y_{apex})$ of the apex point can be automatically determined according to the conditions in Table 2. Let $R_{\mu_e}$ be the set of $< o, (x_{apex}, y_{apex}) >$ tuples where object $o \in A_{n_s} \cup A_{n_e} \cup O_{(n_s,n_{s+1},...,n_e)}$ and $(x_{apex}, y_{apex})$ represents the coordinate of the apex point for the object $o$. Then, for objects $a$, $b$, $c$ in

Table 2
Coordinate of the apex point for object $o \in (A_{n_s} \cup A_{n_e} \cup O_{(n_s, n_{s+1}, \ldots n_e)})$

| Condition | $x_{apex}$ | $y_{apex}$ |
|---|---|---|
| $o \in O_{(n_s, n_{s+1}, \ldots n_e)}$ | $d(n_s, o)$ | 0 |
| $o \in A_{n_s}$ (e.g., $a, b \in A_{n_4}$) | 0 | $d(n_s, o)$ |
| $o \in A_{n_e}$ (e.g., $b, c \in A_{n_5}$) | sequence length | $d(n_e, o)$ |

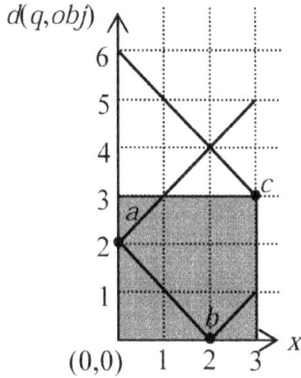

Fig. 6. Constrained query region and qualifying objects $a, b, c$.

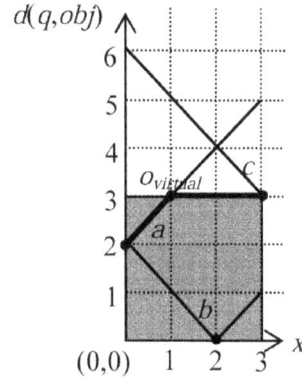

Fig. 7. Marginal objects $a$, $o_{virtual}$, $c$ for [0,1], (1,3), [3,3], respectively.

$A_{n_4}$, $A_{n_5}$, or $O_{(n_4, n_5)}$, $\Phi = \{< a, (0,2) >, < b, (0,2) >, < b, (3,1) >, < c, (3,3) >, < b, (2,0) >\}$.

Observe that object $b$ appears three times in $\Phi$. Two redundant tuples <b,(0,2)> and <b,(3,1)> should be removed as shown in Fig. 5 since <b,(0,2)>, <b,(3,1)>, <b,(2,0)> represent different network distances between $q$ and $b$ and the smallest distance becomes the length of the shortest path between them. After removing the two redundant tuples, $\Phi = \{< a, (0,2) >, < b, (2,0) >, < c, (3,3) >\}$. When $R_{\mu_e}$ includes multiple tuples for the same object, these tuples should be carefully handled using the *cover* relationship [6]. For instance, consider that $\Phi = \{< o_{same}, (0,5) >, < o_{same}, (3,5) >\}$ has two tuples for the same object $o_{same}$ and sequence length is 3. In this case, none of the two tuples is removed.

### 6.2. Decision of valid intervals and their query answer

Figure 6 shows the relationship between qualifying objects and the constrained query region. In this figure, the gray rectangle represents the constrained query region. The query distance and sequence length correspond to the height and width of the rectangle, respectively.

We now elaborate on how to decide the valid intervals and query answer for each valid interval. First, at the start boundary node $n_s$, the $k$-th NN $o_{kth}$ of $q$ is chosen. Unless $o_{kth}$ is found due to a shortage of qualifying objects in $\Phi$, the farthest object $o_{farthest}$ of $q$ at $n_s$ is chosen and is regarded as $o_{kth}$. A virtual object $o_{virtual}$ is introduced to represent the constrained query region and its distance to $q$ is equal to query distance $q.r$. A marginal object is chosen between $o_{kth}$ and $o_{virtual}$. A valid interval has the same marginal object whose distance to $q$ is not smaller than the distance of $q$ to every qualifying object in the valid interval. In other words, If $d(q, o_{kth}) \leqslant d(q, o_{virtual})$, $o_{kth}$ becomes the marginal object. Otherwise, $o_{virtual}$ becomes the marginal object. To determine a split point where query results of contiguous valid intervals become different, we simply keep an eye on the marginal object and note when it changes.

Table 3
Experiment parameter settings

| Parameter | Range |
|---|---|
| Number of objects ($N_{obj}$) | 0.5 k $\sim$ 8 k |
| Distribution of objects ($D_{obj}$) | Skewed/uniform |
| Number of moving queries ($N_{qry}$) | 1 k $\sim$ 100 k |
| Number of requested nearest neighbors ($k$) | 8 $\sim$ 64 |
| Query distance ($r$) | 1 km $\sim$ 8 km |
| Maximum observation distance ($R_{max}$) | 1 km $\sim$ 5 km |
| Query speed ($Q_{sp}$) | 60 km/hr |
| Server cache size/Client cache size | 128MB/32 KB |

Figure 7 identifies marginal objects among objects $a$, $b$, $c$, and $o_{virtual}$. For the interval [0,1], object $a$ becomes the marginal object since object $a$ is $o_{kth}$ for the interval and $d(q,a) \leqslant d(q, o_{virtual})$. However, for interval (1,2], $o_{virtual}$ becomes the marginal object since object $a$ is $o_{kth}$ for interval (1,2] but $d(q, o_{virtual}) < d(q,a)$. Similarly, for interval (2,3), $o_{virtual}$ becomes the marginal object since $d(q, o_{virtual}) < d(q,c)$. Finally, for interval [3,3], object $c$ becomes the marginal object. The bold line segments represent valid intervals [0,1], (1,3), and [3,3] of the marginal objects $a$, $o_{virtual}$, and $c$, respectively.

To produce query result for a valid interval, it is sufficient to find qualifying objects $o$ such that for each object $o \in \Phi$, $d(q,o) \leqslant d(q, o_{marginal})$ where $o_{marginal}$ denotes the marginal object for the interval. Consequently, the query answer $q.A$ for $SQ_{(n_4,n_5)}$ is given as follows: $q.A = \{(I, A_I)|([0,1], \{a,b\}), ((1,3), \{b\}), ([3,3], \{b,c\})\}$ where the first attribute $I$ is a valid interval and the second attribute $A_I$ is the query answer for the valid interval $I$.

It is worth noting that the C$k$NN query can be transformed to either a range query or $k$-NN query depending on the query conditions. In an extreme case, the C$k$NN query becomes a range query when the $k$ value is set to infinity. In contrast, the C$k$NN query becomes a $k$-NN query when the $r$ value is set to infinity.

## 7. Performance study

In this section, we evaluate the performance of DAEMON and GMA [18] under a variety of conditions. Section 7.1 describes experimental settings and Section 7.2 presents experimental results.

### 7.1. Experimental settings

The road map of Stanton County, Kansas in the United States, which contains 8,512 nodes and 9,597 road segments, is used as the road network in our experiment [30]. The road map size is approximately 1,617 km$^2$ (= 49 km $\times$ 33 km). The number of intersection nodes is 1,736 and the number of sequences is 3,153.

Each workload consists of from 1 k to 100 k C$k$NN queries which correspond to mobile clients moving at 60 km/hr for 10 minutes. We simulate moving clients (mobile queries) by using the spatio-temporal data generator described in [2]. In GMA, mobile clients inform the server of their changed locations every second. The initial locations of the queries are randomly distributed around the road network and their next road segments are selected with even probability. The distribution of the objects is either uniform or skewed and the cardinality of the objects varies between 0.5 k and 8 k. In other words, when the object cardinalities are 0.5 k and 8 k, average object densities become 0.3/km$^2$ and 4.9/km$^2$,

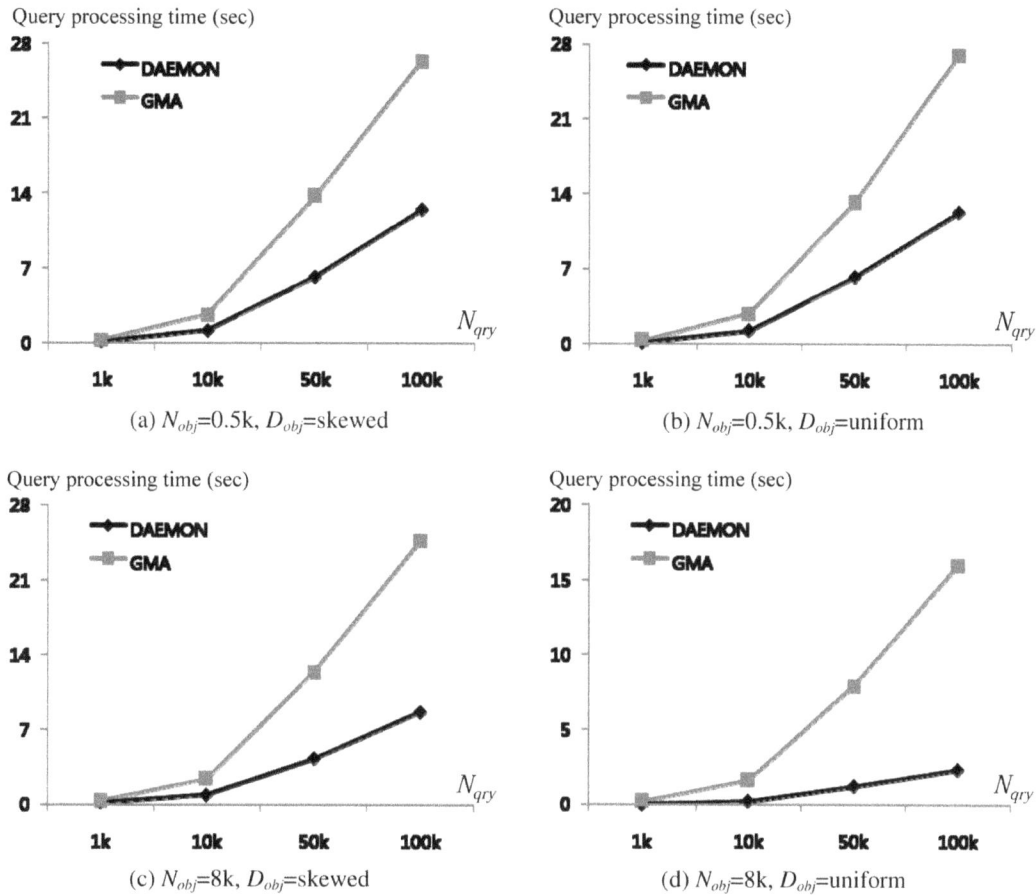

(a) $N_{obj}=0.5k$, $D_{obj}=$skewed

(b) $N_{obj}=0.5k$, $D_{obj}=$uniform

(c) $N_{obj}=8k$, $D_{obj}=$skewed

(d) $N_{obj}=8k$, $D_{obj}=$uniform

Fig. 8. DAEMON vs. GMA in terms of query processing time ($k = [8,16]$, $r = [3$ km, $7$ km]).

respectively. The server cache size is set to 128 MB and the client cache size is set to 32 KB. The former is used to maintain information on condensing nodes while the latter is used to cache server responses for current and previous active sequences. Table 3 summarizes the parameters used in the experiment. All the experiments are conducted on a Pentium 2.8 GHz CPU with 4 GBytes of memory.

We investigate the performance of DAEMON and GMA with two major measures: (1) query processing time and (2) messaging cost. The query processing time is the average of the sum of query processing times at the server and clients while the clients travel along their active sequence. The messaging cost is measured by the size and the number of messages transferred between the clients and server while the clients travel along their active sequence.

### 7.2. Experimental results

Figure 8 shows a performance comparison of DAEMON and GMA using query workloads where the $k$ value of each query is randomly chosen between 8 and 16 and the $r$ value is randomly chosen between 3 km and 7 km. The result trend is so similar that we only show a small selection of these results. On average, the performance of DAEMON is 3.2 times better than that of GMA. The performance gap increases with the $N_{qry}$ value. This is expected because in GMA, the frequent location updates of clients

(a) $N_{obj}$=0.5k, $D_{obj}$=skewed, $N_{qry}$=10k

(b) $N_{obj}$=0.5k, $D_{obj}$=uniform, $N_{qry}$=10k

(c) $N_{obj}$=8k, $D_{obj}$=skewed, $N_{qry}$=10k

(d) $N_{obj}$=8k, $D_{obj}$=uniform, $N_{qry}$=10k

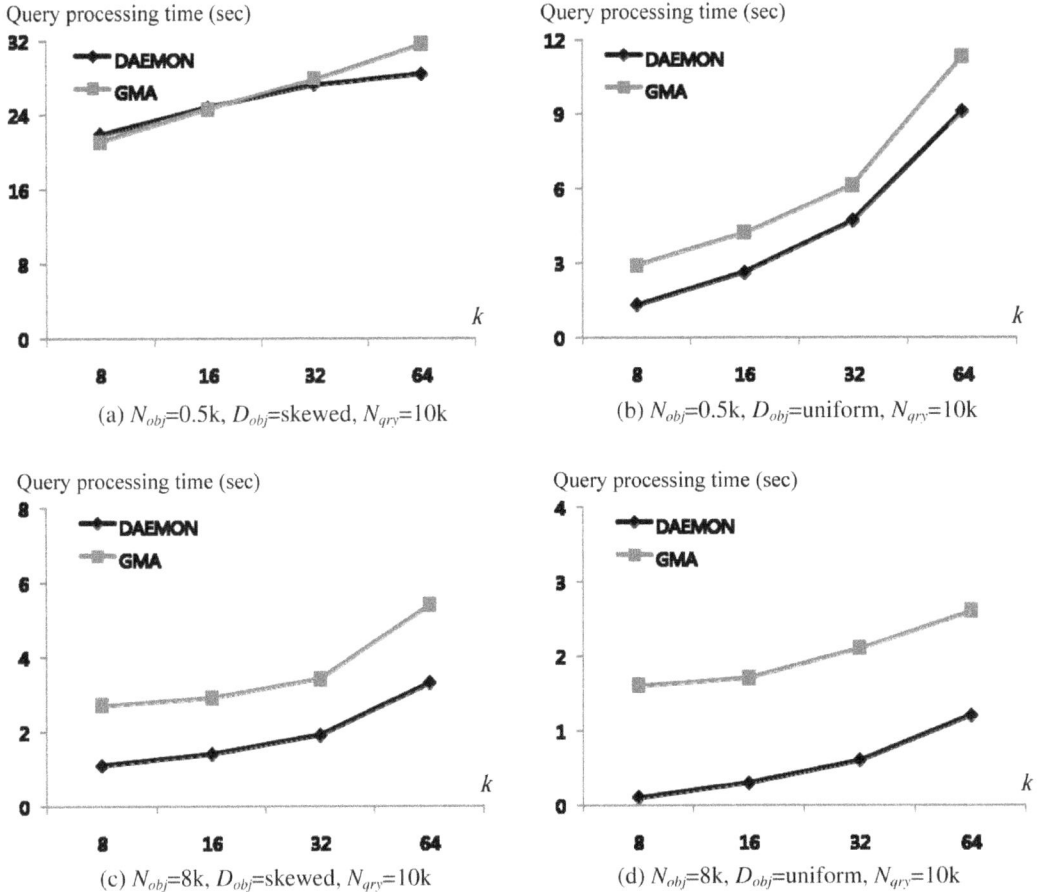

Fig. 9. DAEMON vs. GMA in terms of query processing time ($r = \infty$, $k = [8,64]$).

incur the high wireless communication cost and trigger query reevaluation. In GMA, the client sends location update messages periodically (e.g., every second) to the server. The server re-evaluates the C$k$NN query whenever the movement of the client occurs. The server should provide the updated result to the client if the movement invalidates current query result. Such a situation incurs high computational costs and messaging costs particularly when $N_{obj}$ (e.g., $N_{obj} = 8$ k) is large and $N_{qry}$ (e.g., $N_{qry} = 100$ k) is also large as seen in Figs 8(c) and 8(d).

In DAEMON, the clients do not report their locations and simply ask for the query results at boundary nodes and for objects on the sequence. The query processing is distributed among the clients and the server. Therefore, in DAEMON, large $N_{qry}$ values do not cause sharp performance degradation. It is interesting to observe that Figs 8(a) and 8(b) show very similar results even if the distributions (i.e., uniform and skewed distributions) are quite different. The reason for this is that the query region is limited by query distance $r$ and the object cardinality is very low (in this case $N_{obj} = 500$) so that there is little difference in the search space despite the difference in the distribution of objects. In summary, DAEMON also shows an increase in query processing time due to the increased server load as the $N_{qry}$ value increases, but its performance improves relative to GMA during this increase in load. In both DAEMON and GMA, the processing time at the server accounts for the majority of the total query processing time.

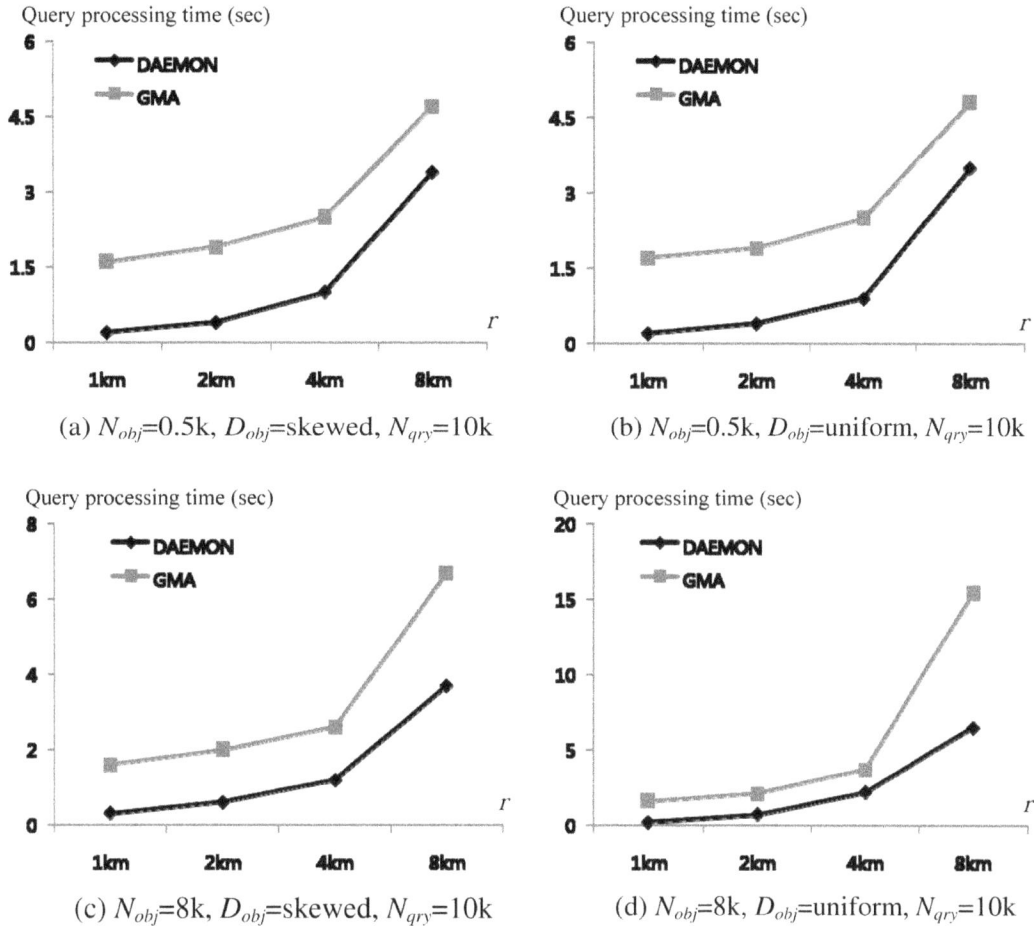

(a) $N_{obj}$=0.5k, $D_{obj}$=skewed, $N_{qry}$=10k

(b) $N_{obj}$=0.5k, $D_{obj}$=uniform, $N_{qry}$=10k

(c) $N_{obj}$=8k, $D_{obj}$=skewed, $N_{qry}$=10k

(d) $N_{obj}$=8k, $D_{obj}$=uniform, $N_{qry}$=10k

Fig. 10. DAEMON vs. GMA in terms of query processing time ($k = \infty$, $r = [1$ km, $8$ km]).

Figure 9 shows a performance comparison of DAEMON and GMA using workloads where the $k$ value of each query varies from 8 to 64 and the $r$ value is set to infinity. These workloads simulate continuous NN queries. The query performance of DAEMON is greatly affected by the density and distribution of objects since the $r$ value is set to infinity. As seen in Fig. 9(a), when a small number of objects are distributed nonuniformly, the search space is so large that the condensing nodes are of little use. On average, the performance of DAEMON is 2.6 times better than that of GMA. The performance gap between DAEMON and GMA typically increases with the $N_{obj}$ value. The number of valid intervals in an active sequence increases with the $N_{obj}$ value. When the object cardinality is high as seen in Figs 9(c) and 9(d), the server of GMA has to evaluate queries whenever the movements of the clients are reported to the server. If necessary, updated results are sent to the clients. However, in DAEMON, each client sends only one request to the server per active sequence and the server evaluates the query only once for every client.

Figure 10 shows a performance comparison between DAEMON and GMA using workloads where the $r$ value of each query varies from 1 km to 8 km and the $k$ value is set to infinity. These workloads simulate continuous range queries. The search space is limited by the query distance $r$ independent of the density and distribution of objects. As a result, DAEMON substantially outperforms GMA due to

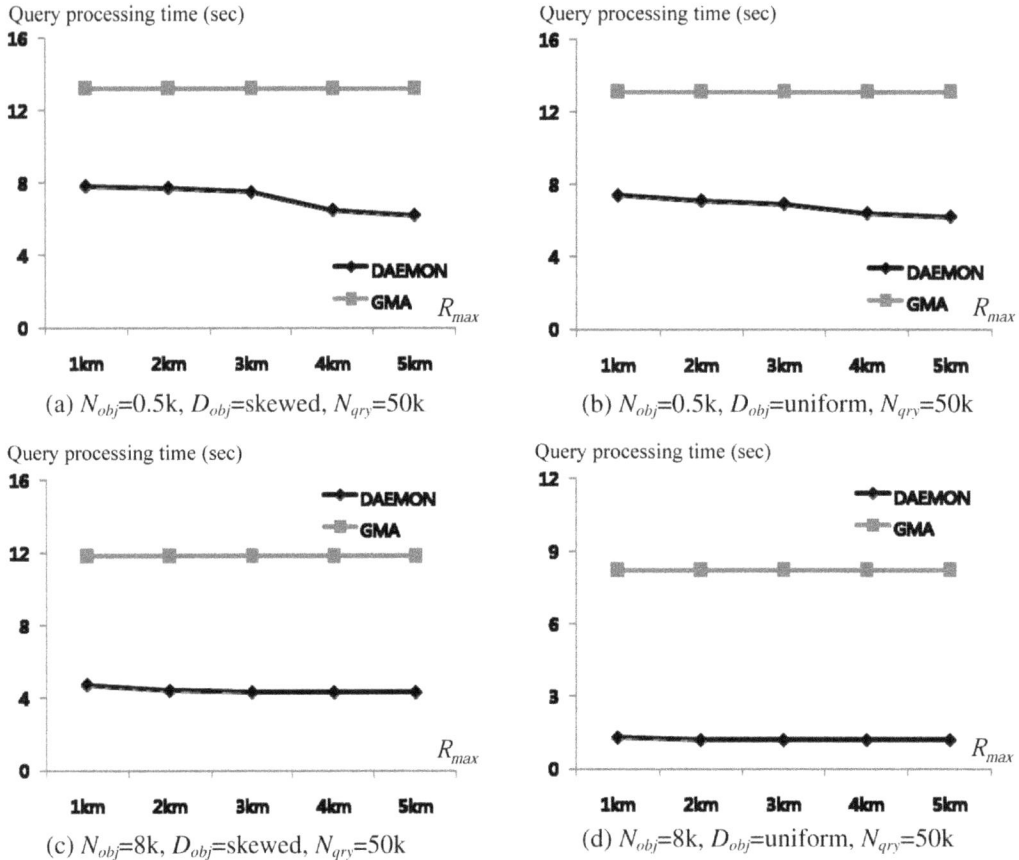

Fig. 11. Query processing time with respect to $R_{max}$ ($k = [8,16]$, $r = [3\ \text{km}, 7\ \text{km}]$).

the use of condensing nodes. On average, the performance of DAEMON is 4.3 times better than that of GMA.

Figure 11 shows the effect of maximum observation distance $R_{\max}$ of condensing nodes on the performance of DAEMON where the $R_{\max}$ value varies from 1 km to 5 km. As the maximum observation distance is larger, DAEMON shows better performance. The reason is that the observation distance of condensing nodes increases with the $R_{\max}$ value. When the cardinality of objects is low, a larger $R_{\max}$ value achieves better performance as seen in Figs 11(a) and 11(b). In Figs 11(c) and 11(d), since the cardinality of objects is high, the change in the $R_{\max}$ value rarely affects the query cost. DAEMON has a trade-off between its query cost and update cost both of which are controlled by the $R_{\max}$ value. As the $R_{\max}$ value increases, the query cost of DAEMON decreases while the update cost increases. To minimize the query cost for a given query distance $r$, it is sufficient for $R_{\max} \geqslant r$.

Figure 12 shows a comparison of messaging costs between DAEMON and GMA using workloads where the $k$ value of each query varies between 8 and 64 and the $r$ value is fixed to 5 km. The numbers in parentheses of these figures refer to the total number of messages transferred between the client and server while the client stays on the active sequence. It is important to note that DAEMON always requires two messages which are the client request and server response messages. In terms of messaging cost, DAEMON achieves a close-to-optimal communication cost.

In GMA, the client reports its location periodically (e.g., every second) and the server evaluates the

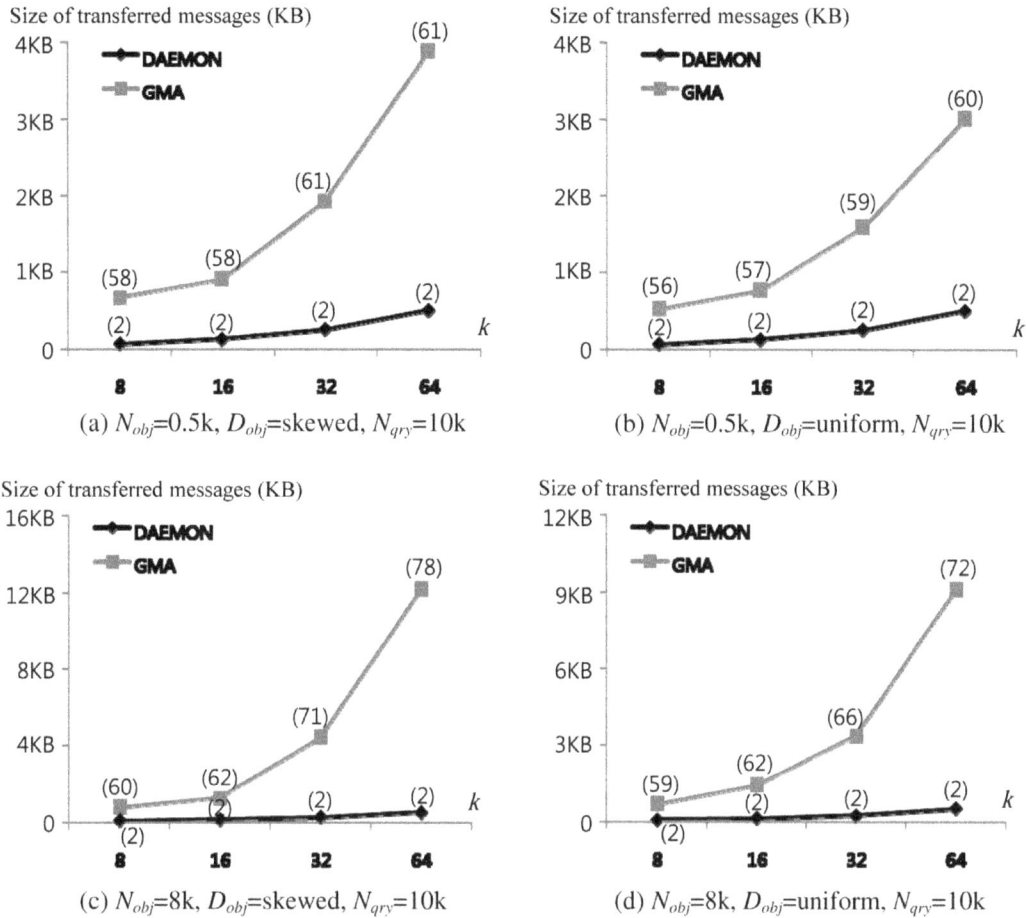

(a) $N_{obj}$=0.5k, $D_{obj}$=skewed, $N_{qry}$=10k

(b) $N_{obj}$=0.5k, $D_{obj}$=uniform, $N_{qry}$=10k

(c) $N_{obj}$=8k, $D_{obj}$=skewed, $N_{qry}$=10k

(d) $N_{obj}$=8k, $D_{obj}$=uniform, $N_{qry}$=10k

Fig. 12. DAEMON vs. GMA in terms of messaging cost ($r = 5$ km, $k = [8,64]$).

query based on the last updated location of the client and informs the client of the changed result if necessary. Specifically, in GMA, when the $k$ values are set to 8, 16, 32, and 64, the ratio of the number of location update messages to the total number of messages are on average 95%, 92%, 87%, and 81%, respectively. For larger $k$ values, the server provides more updated results to the clients. This is because the number of valid intervals in the active sequence increases with the $k$ value. In our road map, the average length of a sequence is about 0.93 km. For example, if a client moves at 60 km/hr, it takes about 55.8 seconds for the client to go through the active sequence. This means that a GMA client sends an average of 55 location update messages to the server while on a particular active sequence.

Figure 13 shows the difference of the size of transferred messages in DAEMON when the client cache is turned on and off. The size of transferred messages is explored under the same conditions as Fig. 8. When the client cache is turned on and the server response for the previous active sequence is cached, the size of transferred messages is about 1.76 times smaller than when the client cache is turned off. This is expected because when using the client cache, the client typically asks for qualifying objects at a boundary node of the active sequence rather than those at both boundary nodes. It is interesting to observe that despite the difference in the cardinality of objects, there is little difference in the size of transferred messages between when $N_{obj} = 8$ k and $D_{obj} = $ skewed and when $N_{obj} = 0.5$ k and $D_{obj} = $ uniform. When $N_{obj} = 8$ k and $D_{obj} = $ skewed, in many cases, the server response includes no

Size of transferred messages (bytes)

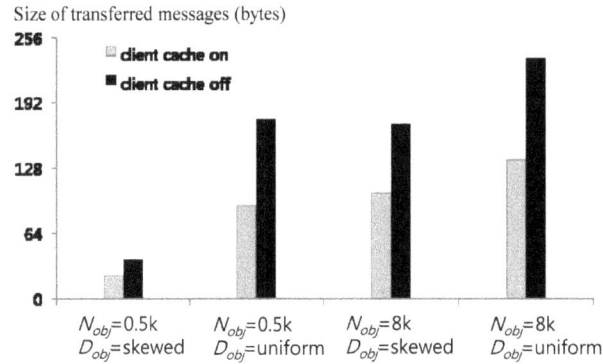

Fig. 13. Client cache on vs. client cache off ($k = [8,16]$, $r = [3$ km, $7$ km]).

qualifying objects or a smaller number of qualifying objects than the requested size due to the skewed distribution of objects.

## 8. Conclusions

We propose a distributed and scalable strategy, DAEMON, for the continuous monitoring of C$k$NN queries on road networks. In DAEMON, the server efficiently evaluates C$k$NN queries issued at boundary nodes and quickly retrieves the objects on the sequence using dedicated data structures: the condensing nodes and sequences. Each client sends only one request to the server while traveling on consecutive road segments between adjacent intersections and then generates query result for these road segments using the server response. Unlike the centralized solutions, in DAEMON, the clients do not send location update messages to the server, which reduces the communication cost and alleviates concerns of location privacy. Our distributed processing of C$k$NN queries significantly decreases the computation cost of the server and the communication cost between clients and the server while placing only a small processing burden on the mobile clients. Our extensive experimental results confirm that DAEMON clearly outperforms its rival in terms of query processing time while achieving close-to-optimal communication costs.

## Acknowledgements

This research was supported by Basic Science Research Program through the National Research Foundation of Korea (NRF) funded by the Ministry of Education, Science and Technology (2010-0013487).

## References

[1]   J. Bao, C. Chow, M. Mokbel and W. Ku, Efficient evaluation of $k$-Range nearest neighbor queries in road networks, *Mobile Data Management* (2010), 115–124.
[2]   T. Brinkhoff, A framework for generating network-based moving objects, *GeoInformatica* **6**(2) (2002), 153–180.
[3]   Y. Cai, K. Hua and G. Cao, Processing range-monitoring queries on heterogeneous mobile objects, *Mobile Data Management* (2004), 27–38.
[4]   M. Cheema, L. Brankovic, X. Lin, W. Zhang and W. Wang, Continuous monitoring of distance-based range queries, *IEEE Trans Knowl Data Eng* **23**(8) (2011), 1182–1199.

[5]     Z. Chen, H. Shen, X. Zhou and J. Yu, Monitoring path nearest neighbor in road networks, *SIGMOD Conference* (2009), 591–602.

[6]     H. Cho and C. Chung, An efficient scalable approach to CNN Queries in a road network, *VLDB* (2005), 865–876.

[7]     T. Delot, S. Ilarri, N. Cenerario and T. Hien, Event sharing in vehicular networks using geographic vectors and maps, *Mobile Information Systems* **7**(1) (2011), 21–44.

[8]     H. Ferhatosmanoglu, I. Stanoi, D. Agrawal and A. Abbadi, Constrained nearest neighbor queries, *SSTD* (2001), 257–278.

[9]     Y. Gao, G. Chen, Q. Li, C. Li and C. Chen, Constrained k-Nearest neighbor query processing over moving object trajectories, *DASFAA* (2008), 635–643.

[10]    B. Gedik and L. Liu, MobiEyes: A Distributed location monitoring service using moving location queries, *IEEE Trans Mob Comput* **5**(10) (2006), 1384–1402.

[11]    N. Ghadiri, A. Dastjerdi, N. Aghaee and M. Nematbakhsh, Optimizing the performance and robustness of type-2 fuzzy group nearest-neighbor queries, *Mobile Information Systems* **7**(2) (2011), 123–145.

[12]    M. Hasan, M. Cheema, W. Qu and X. Lin, Efficient algorithms to monitor continuous constrained $k$ nearest neighbor queries, *DASFAA* (1) (2010), 233–249.

[13]    H. Hu, J. Xu and D. Lee, PAM: An efficient privacy-aware monitoring framework for continuously moving objects, *IEEE Trans Knowl Data Eng* **22**(3) (2010), 404–419.

[14]    S. Ilarri, E. Mena and A. Illarramendi, Location-dependent query processing: Where we are where we are heading, *ACM Comput Surv* **42**(3) (2010).

[15]    M. Mokbel, X. Xiong and W. Aref, SINA: Scalable incremental processing of continuous queries in spatio-temporal databases, *SIGMOD Conference* (2004), 623–634.

[16]    D. Moon, B. Park, Y. Chung and J. Park, Recovery of flash memories for reliable mobile storages, *Mobile Information Systems* **6**(2) (2010), 177–191.

[17]    F. Morvan and A. Hameurlain, A mobile relational algebra, *Mobile Information Systems* **7**(1) (2011), 1–20.

[18]    K. Mouratidis, M. Yiu, D. Papadias and N. Mamoulis, Continuous nearest neighbor monitoring in road networks, *VLDB* (2006), 43–54.

[19]    K. Mouratidis and M. Yiu, Anonymous query processing in road networks, *IEEE Trans Knowl Data Eng* **22**(1) (2010), 2–15.

[20]    D. Papadias, J. Zhang, N. Mamoulis and Y. Tao, Query processing in spatial network databases, *VLDB* (2003), 802–813.

[21]    D. Stojanovic, A. Papadopoulos, B. Predic, S. Kajan and A. Nanopoulos, Continuous range monitoring of mobile objects in road networks, *Data Knowl Eng* **64**(1) (2008), 77–100.

[22]    W. Wu, W. Guo and K. Tan, Distributed processing of moving K-Nearest-Neighbor query on moving objects, *ICDE* (2007), 1116–1125.

[23]    K. Xuan, G. Zhao, D. Taniar, J. Rahayu, M. Safar and B. Srinivasan, Voronoi-based range and continuous range query processing in mobile databases, *J Comput Syst Sci* **77**(4) (2011), 637–651.

[24]    K. Xuan, G. Zhao, D. Taniar, M. Safar and B. Srinivasan, Constrained range search query processing on road networks. Concurrency and Computation, *Practice and Experience* **23**(5) (2011), 491–504.

[25]    K. Xuan, G. Zhao, D. Taniar, M. Safar and B. Srinivasan, Voronoi-based multi-level range search in mobile navigation, *Multimedia Tools Appl* **53**(2) (2011), 459–479.

[26]    K. Xuan, G. Zhao, D. Taniar and B. Srinivasan, Continuous Range search query processing in mobile navigation, *ICPADS* (2008), 361–368.

[27]    G. Zhao, K. Xuan, W. Rahayu, D. Taniar, M. Safar, M. Gavrilova and B. Srinivasan, Voronoi-based continuous k nearest neighbor search in mobile navigation, *IEEE Transactions on Industrial Electronics* **58**(6) (2011), 2247–2257.

[28]    G. Zhao, K. Xuan and D. Taniar, Path kNN query processing in Mobile Systems, *IEEE Transactions on Industrial Electronics*, to appear (DOI: 10.1109/TIE.2011.2167113).

[29]    G. Zhao, K. Xuan, D. Taniar and B. Srinivasan, LookAhead continuous KNN mobile query processing, *Comput Syst Sci Eng* **25**(3) (2010), 205–217.

[30]    U.S. Census Bureau – TIGER/Line http://www.census.gov/geo/www/tiger/.

**Hyung-Ju Cho** received the B.S. and M.S. degrees in Computer Engineering from Seoul National University in February 1997 and in February 1999, respectively, and the Ph.D. degree in Computer Science from KAIST, in August 2005. He is currently a research assistant professor at the Dept. of Computer Engineering, Ajou University, South Korea. His current research interests include moving objects databases and query processing in mobile peer-to-peer networks.

**Seung-Kwon Choe** got his BS degree in Computer Engineering at Ajou University, South Korea in February 2010. Currently, he pursues the MS degree in Computer Engineering at Ajou University. His research area includes flash memory storages and moving objects databases.

**Tae-Sun Chung** received the B.S. degree in Computer Science from KAIST, in February 1995, and the M.S. and Ph.D. degree in Computer Science from Seoul National University, in February 1997 and August 2002, respectively. He is currently an associate professor at School of Information and Computer Engineering at Ajou University. His current research interests include flash memory storages, XML databases, and database systems.

# Permissions

The contributors of this book come from diverse backgrounds, making this book a truly international effort. This book will bring forth new frontiers with its revolutionizing research information and detailed analysis of the nascent developments around the world.

We would like to thank all the contributing authors for lending their expertise to make the book truly unique. They have played a crucial role in the development of this book. Without their invaluable contributions this book wouldn't have been possible. They have made vital efforts to compile up to date information on the varied aspects of this subject to make this book a valuable addition to the collection of many professionals and students.

This book was conceptualized with the vision of imparting up-to-date information and advanced data in this field. To ensure the same, a matchless editorial board was set up. Every individual on the board went through rigorous rounds of assessment to prove their worth. After which they invested a large part of their time researching and compiling the most relevant data for our readers.

The editorial board has been involved in producing this book since its inception. They have spent rigorous hours researching and exploring the diverse topics which have resulted in the successful publishing of this book. They have passed on their knowledge of decades through this book. To expedite this challenging task, the publisher supported the team at every step. A small team of assistant editors was also appointed to further simplify the editing procedure and attain best results for the readers.

Apart from the editorial board, the designing team has also invested a significant amount of their time in understanding the subject and creating the most relevant covers. They scrutinized every image to scout for the most suitable representation of the subject and create an appropriate cover for the book.

The publishing team has been an ardent support to the editorial, designing and production team. Their endless efforts to recruit the best for this project, has resulted in the accomplishment of this book. They are a veteran in the field of academics and their pool of knowledge is as vast as their experience in printing. Their expertise and guidance has proved useful at every step. Their uncompromising quality standards have made this book an exceptional effort. Their encouragement from time to time has been an inspiration for everyone.

The publisher and the editorial board hope that this book will prove to be a valuable piece of knowledge for researchers, students, practitioners and scholars across the globe.

# List of Contributors

**Sazia Parvin**
Digital Ecosystems and Business Intelligence Institute, Curtin University, Perth, Australia

**Farookh Khadeer Hussain**
School of Software, Faculty of Engineering and Information Technology, University of Technology, Sydney, Australia

**Sohrab Ali**
The People's University of Bangladesh, Dhaka, Bangladesh

**Keum-Sung Hwang**
Department of Computer Science, Yonsei University, Seoul, Korea

**Sung-Bae Cho**
Department of Computer Science, Yonsei University, Seoul, Korea

**Minho Kim**
School of Electrical Engineering and Computer Science, Seoul National University, Seoul 151-742, Republic of Korea

**Eun-Chan Park**
Department of Information and Communication Engineering, Dongguk University, Seoul 100-715, Republic of Korea

**Chong-Ho Choi**
School of Electrical Engineering and Computer Science, Seoul National University, Seoul 151-742, Republic of Korea

**Radu Ioan Ciobanu**
Faculty of Automatic Control and Computers, University Politehnica of Bucharest, Romania

**Ciprian Dobre**
Faculty of Automatic Control and Computers, University Politehnica of Bucharest, Romania

**Valentin Cristea**
Faculty of Automatic Control and Computers, University Politehnica of Bucharest, Romania

**Florin Pop**
Faculty of Automatic Control and Computers, University Politehnica of Bucharest, Romania

**Fatos Xhafa**
Universitat Politecnica de Catalunya, Barcelona, Spain

**YileiWang**
School of Computer Science and Technology, Shandong University, Jinan 250101, China
School of Information and Electrical Engineering, Ludong University, Yantai 264025, China

**Chuan Zhao**
School of Computer Science and Technology, Shandong University, Jinan 250101, China

**Qiuliang Xu**
School of Computer Science and Technology, Shandong University, Jinan 250101, China

**Zhihua Zheng**
School of Information Science and Engineering, Shandong Normal University, Jinan 250014, China

**Zhenhua Chen**
School of Computer Science, Shaanxi Normal University, Xi'an 710062, China

**Zhe Liu**
Laboratory of Algorithmics, Cryptology and Security (LACS), 1359 Luxembourg, Luxembourg

**Yunchuan Sun**
Business School, Beijing Normal University, Beijing, China

**Hongli Yan**
Business School, Beijing Normal University, Beijing, China

**Cheng Lu**
College of Information Science and Technology, Beijing Normal University, Beijing, China

**Rongfang Bie**
College of Information Science and Technology, Beijing Normal University, Beijing, China

**Zhangbing Zhou**
School of Information Engineering, China University of Geosciences, Beijing, China
Computer Science Department, Institute Mines-TELECOM/TELECOM SudParis, Paris, France

**Cang-hong Jin**
College of Computer Science and Technology, Zhejiang University, Hangzhou 310027, China
Department of Computer Science and Engineering, Zhejiang University City College, Hangzhou 310015, China

**Ze-min Liu**
College of Computer Science and Technology, Zhejiang University, Hangzhou 310027, China

**Ming-hui Wu**
Department of Computer Science and Engineering, Zhejiang University City College, Hangzhou 310015, China

**Jing Ying**
College of Computer Science and Technology, Zhejiang University, Hangzhou 310027, China

**Kam-Yiu Lam**
Department of Computer Science, City University of Hong Kong, Kowloon Tong, Hong Kong

**Joseph Kee Yin Ng**
Department of Computer Science, Hong Kong Baptist University, Kowloon Tong, Hong Kong

**JiantaoWang**
Department of Computer Science, Hong Kong Baptist University, Kowloon Tong, Hong Kong

**Calvin Ho Chuen Kam**
Department of Computer Science, Hong Kong Baptist University, Kowloon Tong, Hong Kong

**NelsonWai-Hung Tsang**
Department of Computer Science, City University of Hong Kong, Kowloon Tong, Hong Kong

**Antonio J. Jara**
Department of Information and Communication Engineering, University of Murcia, Murcia, Spain

**Miguel A. Zamora**
Department of Information and Communication Engineering, University of Murcia, Murcia, Spain

**Antonio Skarmeta**
Department of Information and Communication Engineering, University of Murcia, Murcia, Spain

**Hui-Huang Hsu**
Department of Computer Science and Information Engineering, Tamkang University, 151 Ying-Chuan Rd., Tamsui, New Taipei City, Taiwan

**Cheng-Ning Lee**
Department of Computer Science and Information Engineering, Tamkang University, 151 Ying-Chuan Rd., Tamsui, New Taipei City, Taiwan

**Jason C. Hung**
Department of Information Management, Overseas Chinese University, 100 Chiao Kwang Rd., Taichung, Taiwan

**Timothy K. Shih**
Department of Computer Science and Information Engineering, National Central University, 300 Jhongda Rd., Jhongli, Taoyuan, Taiwan

**Hyung-Ju Cho**
Department of Computer Engineering, Ajou University, Gyeonggi-Do, South Korea

**Seung-Kwon Choe**
Department of Computer Engineering, Ajou University, Gyeonggi-Do, South Korea

**Tae-Sun Chung**
Department of Computer Engineering, Ajou University, Gyeonggi-Do, South Korea

**ShangguangWang**
State Key Laboratory of Networking and Switching Technology, Beijing University of Posts and Telecommunications, Haidian, Beijing 100876, China

**Lei Sun**
State Key Laboratory of Networking and Switching Technology, Beijing University of Posts and Telecommunications, Haidian, Beijing 100876, China

**Qibo Sun**
State Key Laboratory of Networking and Switching Technology, Beijing University of Posts and Telecommunications, Haidian, Beijing 100876, China

**Xuyan Li**
Basic Research Service Ministry of Science and Technology the People's Republic of China, Haidian, Beijing 100862, China

**Fangchun Yang**
State Key Laboratory of Networking and Switching Technology, Beijing University of Posts and Telecommunications, Haidian, Beijing 100876, China

www.ingramcontent.com/pod-product-compliance
Lightning Source LLC
Chambersburg PA
CBHW070155240326

41458CB00126B/5158